Solutions Manual and Computer Programs for

Physical and Computational Aspects of Convective Heat Transfer

Tuncer Cebeci

Solutions Manual and Computer Programs for **Physical and Computational Aspects of Convective Heat Transfer**

by T. Cebeci and P. Bradshaw

With 30 Illustrations

Springer-Verlag
New York Berlin Heidelberg
London Paris Tokyo

Tuncer Cebeci
Douglas Aircraft Company
Long Beach, California 90846
U.S.A.

and

Department of Aerospace Engineering
California State University, Long Beach
Long Beach, California 90840
U.S.A.

Text prepared by the author in camera-ready form.

Printed in the United States of America.

9 8 7 6 5 4 3 2 1

ISBN 0-387-96825-3 Springer-Verlag New York Berlin Heidelberg
ISBN 3-540-96825-3 Springer-Verlag Berlin Heidelberg New York

Preface

This manual deals with the problems and the computer programs described in the book Physical and Computational Aspects of Convective Heat Transfer and is prepared in two parts. The first part provides solutions of the problems of Chapters 1 through 9 of the book, and vary in that in some cases a short answer suffices, while in others the use of a computer program is required. No solutions are provided for the problems of Chapter 10, which are intended to stimulate the reader rather than to test his or her knowledge. The second part presents a description of the computer programs of Chapters 13 and 14 of the book, which allow the user to solve problems of internal flows, compressible external boundary layers, wall jets, free jets, and mixing layers. The computer programs of Parts 1 and 2 are included in one 5-1/4" double-sided diskette which may be obtained from the author, Aerospace Engineering Department, California State University, Long Beach, Long Beach, CA 90840 with a remittance of $50.

I am grateful to my colleagues, especially to Drs. K.C. Chang and N. Alemdaroglu, and to my graduate student, David Egan, for their assistance in the formulation of some of the solutions, to Nancy Barela for all her patience and excellent skill in the typing of this manual, and to Professors Peter Bradshaw and Jim Whitelaw for their continuing interest and advice. I will be pleased to learn of alternative procedures for the solution of the problems.

May 1988 Tuncer Cebeci

Palm Springs

Table of Contents

PART 1

SOLUTIONS TO PROBLEMS OF CHAPTERS 1 TO 9

Chapter 1

Introduction

1.1 a. With ρ = 879 kg/m^3, μ = 3.3 x 10^{-4} kg/ms, κ = 66.5 x 10^{-6}m^2/s, Pr = $\mu/\rho\kappa$ = 3.3 x 10^{-4}/(879.0 x 66.5 x 10^{-6}) = 0.0056.

b. With c_p = 0.2415 Btu/lb°F, μ = 5.193 x 10^{-2}lb/ft h, k = 1.806 x 10^{-2} Btu/h ft °F, Pr = $\mu c_p/k$ = 5.193 x 10^{-2} x 0.2415/ (1.806 x 10^{-2}) = 0.694.

c. With ρ = 900 kg/m^3, c_p = 1902 kJ/kg K, ν = 9 x 10^{-4}m^2/s, k = 0.143 W/m K, Pr = $\mu c_p/k$ = $\rho\nu c_p/k$ = 900.0 x (9 x 10^{-4}) x 1.902 x 10^3/0.143 = 10773.6.

1.2 a. With u_∞ = 100 ft/s, L = 0.5 ft and ν = 2.15 x 10^{-4} ft^2/s for air at T = 150°F and p = 14.7 lb/in^2, R_L = $u_\infty L/\nu$ = 100 x 0.5/(2.15 x 10^{-4}) = 2.33 x 10^5.

b. With ν = 0.925x10^{-6}m^2/s for water at T = 25°C, p = 14.7 lb/in^2, u_m = 12m/s, d = 0.25m, R_d = $u_m d/\nu$ = 12.0 x 0.25/(0.925 x 10^{-6}) = 3.24 x 10^6.

c. With u_e = 10m/s, x = 1m and ν = 5.0 x 10^{-4}m^2/s for glycerine at T = 3°C, R_x = $u_e x/\nu$ = 10.0 x 1.0/(5.0 x 10^{-4}) = 2.0 x 10^4.

1.3 a. With u_e = 100 ft/s and ρ = 0.065 lbm/ft^3 for air at T = 150°F and p = 14.7 lb/in^2, τ_w = ρu_e^2 0.332 $(R_x)^{-1/2}$ = 0.065 x 100^2 x 0.332/$\sqrt{2.33 \times 10^5}$ = 0.447 lbm/ft s^2.

b. Since Pr for air is nearly unity, Reynolds analogy should be a good approximation. Then from $S_t/(c_f/2) = 1.0 = \dot{q}_w/[c_f/2\ \rho c_p(T_w - T_e)u_e]$, $\dot{q}_w = 0.332/\sqrt{2.33 \times 10^5} \times 0.065 \times 0.2410 \times 10 \times 100 = 0.0108$ Btu/ft^2 s.

Chapter 2

Conservation Equations for Mass, Momentum and Energy

2.1 a. The derivation of the equation is straightforward; there is no need to elaborate on the details of the algebra.

b. Eq. (p2.1) may be rewritten as

$$\frac{\partial \rho}{\partial t} + u\frac{\partial \rho}{\partial x} + v\frac{\partial \rho}{\partial y} + w\frac{\partial \rho}{\partial z} + \rho\left(\frac{\partial u}{\partial x} + \frac{\partial v}{\partial y} + \frac{\partial w}{\partial z}\right) = 0 \qquad (1)$$

For constant-density flow, $\frac{\partial \rho}{\partial t} = \frac{\partial \rho}{\partial x} = \frac{\partial \rho}{\partial y} = \frac{\partial \rho}{\partial z} = 0$

Hence $\frac{\partial u}{\partial x} + \frac{\partial v}{\partial y} + \frac{\partial w}{\partial z} = 0$ (2)

For steady but variable-density flows, $\partial\rho/\partial t = 0$, and (P2.1) becomes $\partial/\partial x\,(\rho u) + \partial/\partial y\,(\rho v) + \partial/\partial z\,(\rho w) = 0$ (3)

c. Since by definition

$\frac{d\rho}{dt} \equiv \frac{\partial \rho}{\partial t} + u\frac{\partial \rho}{\partial x} + v\frac{\partial \rho}{\partial y} + w\frac{\partial \rho}{\partial z}$, (1) may be written as

$\frac{d\rho}{dt} = \frac{\partial \rho}{\partial t} + u\frac{\partial \rho}{\partial x} + v\frac{\partial \rho}{\partial y} + w\frac{\partial \rho}{\partial z} = -\rho\left(\frac{\partial u}{\partial x} + \frac{\partial v}{\partial y} + \frac{\partial w}{\partial z}\right)$ or as

$\frac{d\rho}{dt} + \rho\nabla \cdot \vec{v} = 0$, which is, by definition, the transport equation for ρ.

2.3 Suppose that in a short time dt, a fluid element moves from p to p', as shown in the figure below. Then a quantity - such as temperature, say - that has the value ϕ at location p and time t will have the value

$$\phi + \frac{\partial \phi}{\partial x}dx + \frac{\partial \phi}{\partial y}dy$$

at point p' and time t, and the value

$$\phi + \frac{\partial\phi}{\partial t}\,dt + \frac{\partial\phi}{\partial x}\,dx + \frac{\partial\phi}{\partial y}\,dy \equiv \phi', \quad \text{say}$$

at point p' and time t + dt (where $\partial\phi/\partial t$ is the time derivative at a fixed point, strictly at p'). Since $dx = u\,dt$, $dy = v\,dt$ we can write

$$\phi' = \phi + \left(\frac{\partial\phi}{\partial t} + u\,\frac{\partial\phi}{\partial x} + v\,\frac{\partial\phi}{\partial y}\right)dt$$

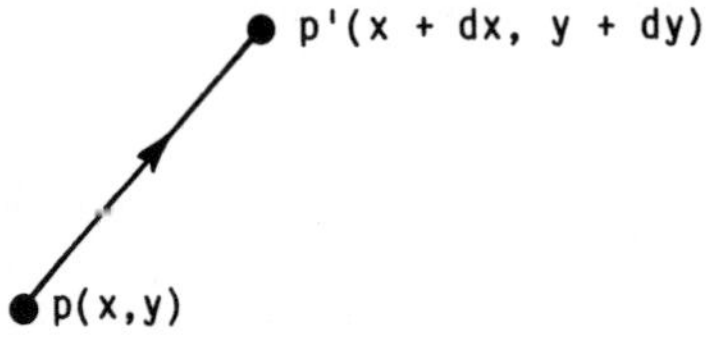

2.6 Use the definition of the substantial derivative, Eq. (2.10), multiply Eq. (2.9) with u, Eq. (2.12) with v, and add the resulting expressions to obtain the desired equation.

2.7 Note that the total-energy equation for two-dimensional flow is given by (2.24). The term, $d(p/\rho)/dt$ can be expanded as

$$\frac{d(p/\rho)}{dt} = \frac{1}{\rho}\frac{dp}{dt} - \frac{p}{\rho^2}\frac{d\rho}{dt} = \frac{1}{\rho}\frac{dp}{dt} + \frac{p}{\rho}\left(\frac{\partial u}{\partial x} + \frac{\partial v}{\partial y}\right)$$

$$= \frac{1}{\rho}\left(u\,\frac{\partial p}{\partial x} + v\,\frac{\partial p}{\partial y}\right) + \frac{p}{\rho}\left(\frac{\partial u}{\partial x} + \frac{\partial v}{\partial y}\right) = \frac{1}{\rho}\left[\frac{\partial}{\partial x}(pu) + \frac{\partial}{\partial y}(pv)\right]$$

and is equal to the second term on the right-hand side of (2.24). Thus the total enthalpy equation can be written as

$$\frac{d}{dt}\left(e + \frac{p}{\rho} + \frac{1}{2}(u^2 + v^2)\right] \equiv \frac{dH}{dt} = -\frac{1}{\rho}\left(\frac{\partial\dot{q}_x}{\partial x} + \frac{\partial\dot{q}_y}{\partial y}\right) + \frac{1}{\rho}\frac{\partial}{\partial x_j}(\sigma_{ij}u_i)$$
$$+ uf_x + vf_y$$

Dealing with total enthalpy instead of total energy has the same advantage as dealing with (static) enthalpy rather than internal energy: the pressure does not appear explicitly.

2.8 a. $\bar{u} = a$, $u' = b\sin\omega t$, $\bar{T} = c$, $T' = d\sin(\omega t - \phi)$, $\overline{u'^2} = b^2/2$, $\overline{T'^2} = d^2/2$, $\overline{u'T'} = bd/2\cos\phi$ (expand $\sin(\omega t - \phi) = \sin\omega t\cos\phi - \sin\phi\cos\omega t$ and note that the average of $\sin\omega t\cos\omega t$ is zero.)

b. $\bar{u} = a + b/2$, $u' = b(\sin^2\omega t - 1/2)$, $\bar{T} = c + d/2$
$T' = d[\sin^2(\omega t - \phi) - 1/2]$, $\overline{u'^2} = b^2/8$, $\overline{T'^2} = d^2/8$
$\overline{u'T'} = (db/8)\cos\phi$ (note that the average of $\sin^2\omega t$ is 1/2)

2.9 Analogous to Eq. (2.43), we write the y component of the momentum equation as

$$(\rho u + \overline{\rho' u'}) \frac{\partial v}{\partial x} + (\rho v + \overline{\rho' v'}) \frac{\partial v}{\partial y} = - \frac{\partial p}{\partial y} + \frac{\partial \sigma_{yx}}{\partial x} + \frac{\partial \sigma_{yy}}{\partial y} + \rho \bar{f}_y + \overline{\rho' f'_y}$$

$$- \{ \frac{\partial}{\partial x} (\rho \overline{u'v'} + \overline{\rho' u' v'} + u \overline{\rho' v'}) \}$$

$$- \{ \frac{\partial}{\partial y} (\rho \overline{v'^2} + \overline{\rho' v'^2} + v \overline{\rho' v'}) \}$$

For constant density all the terms containing ρ' drop out and the resulting equation becomes the desired equation after both sides are divided by ρ.

Chapter 3

Boundary-Layer Equations

3.1 An average value of $\partial u/\partial y$ for a boundary-layer thickness is u_e/δ (u rises from zero at $y = 0$ to u_e at $y = \delta$); an average value of $\partial^2 u/\partial y^2$ can be derived by a similar argument: $\partial u/\partial y$ changes from a value of order u_e/δ at $y = 0$ to zero at $y = \delta$, so an average value of $\partial(\partial u/\partial y)/\partial y \equiv \partial^2 u/\partial y^2$ is of order u_e/δ^2. The velocity at a fixed distance y from the surface, where y is smaller than the value of δ at the streamwise position considered, falls from u_e at the boundary-layer origin, $x = 0$, to a value rather smaller than u_e at the streamwise position considered. Therefore a typical value of $\partial u/\partial x$ is of order u_e/x or, better, $(u_e/\delta)d\delta/dx$ since if the boundary layer really grew linearly at the rate of $d\delta/dx$, its origin would be at a distance $\delta/(d\delta/dx)$ upstream. As before, the argument can be repeated to give a typical value of $\partial^2 u/\partial x^2$ as of order $(u_e/\delta^2)\,(d\delta/dx)^2$, so that finally

$$\frac{\partial^2 u/\partial x^2}{\partial^2 u/\partial y^2} \quad \text{is of order } (d\delta/dx)^2$$

3.2 $V(\frac{\partial p}{\partial s})_\psi = (u\,\frac{\partial}{\partial x} + v\,\frac{\partial}{\partial y})p$ where V is the resultant velocity

$(u^2 + v^2)^{1/2}$, therefore

$$V\,(\frac{\partial p}{\partial s})_\psi = u\,(\frac{\partial p}{\partial x} + \rho u\,\frac{\partial u}{\partial x} + \rho v\,\frac{\partial u}{\partial y}) + v\,(\frac{\partial p}{\partial y} + \rho u\,\frac{\partial v}{\partial x} + \rho v\,\frac{\partial v}{\partial y})$$

$$= u\,(\frac{\partial \sigma_{xx}}{\partial x} + \frac{\partial \sigma_{xy}}{\partial y}) + v\,(\frac{\partial \sigma_{xy}}{\partial x} + \frac{\partial \sigma_{yy}}{\partial y})$$

or $$(\frac{\partial p}{\partial s})_\psi = \frac{u}{V}\,(\frac{\partial \sigma_{xx}}{\partial x} + \frac{\partial \sigma_{xy}}{\partial y}) + \frac{v}{V}\,(\frac{\partial \sigma_{xy}}{\partial x} + \frac{\partial \sigma_{yy}}{\partial y})$$

Now

$$\frac{u}{V} = \frac{u}{(u^2 + v^2)^{1/2}} = (1 + \frac{1}{2}\,\frac{v^2}{u^2} + \ldots) = [1 + O(\frac{\delta}{x})^2]$$

$$\frac{\partial\sigma_{xx}}{\partial x} = \frac{\sigma_{xx}}{\sigma_{yy}}\frac{\partial\sigma_{xy}}{\partial y}\,O(\frac{\delta}{x}), \qquad \frac{v}{V} = \frac{v}{u}\,[1 + O(\frac{\delta}{x})^2] = O(\frac{\delta}{x})$$

$$\frac{\partial\sigma_{xy}}{\partial x} = \frac{\partial\sigma_{xy}}{\partial y}\,O(\frac{\delta}{x}), \qquad \frac{\partial\sigma_{yy}}{\partial y} = O(\frac{\sigma_{yy}}{\sigma_{xy}})\,\frac{\partial\sigma_{xy}}{\partial y}$$

The error in replacing $u/V\,(\partial\sigma_{yy}/\partial y)$ by $(\partial\sigma_{xy}/\partial y)$ is $O(\delta/x)^2$ while the terms in σ_{xx} and σ_{yy} are at most of order δ/x times the main terms (if all σ stresses are of the same order, as in turbulent flow: in laminar flow σ_{xx}, $\sigma_{yy} << \sigma_{xy}$ and all neglected terms are $O(\delta/x)^2$).

3.3 Conservation of mass for incompressible, steady flow with no sources, sinks within the control volume is

(mass flow in) - (mass flow out) = 0

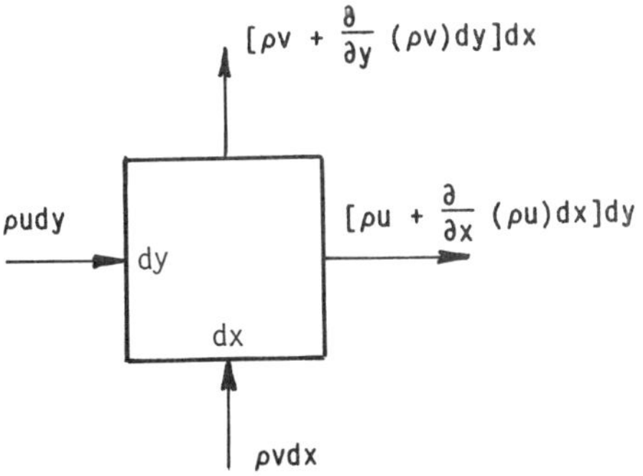

$$\rho u dy + \rho v dx - [\rho u + \frac{\partial}{\partial x}(\rho u)dx]dy - [\rho v + \frac{\partial}{\partial y}(\rho v)dy]dx = 0$$

or $\frac{\partial u}{\partial x} + \frac{\partial v}{\partial y} = 0$

x-momentum equation: Conservation of momentum

Rate of change of momentum in x-direction = $\Sigma(F_x)_{net}$

$\Sigma(F_x)_{net} = (f_{pressure} + F_{viscous} + F_{body})_{x\text{-direction}}$

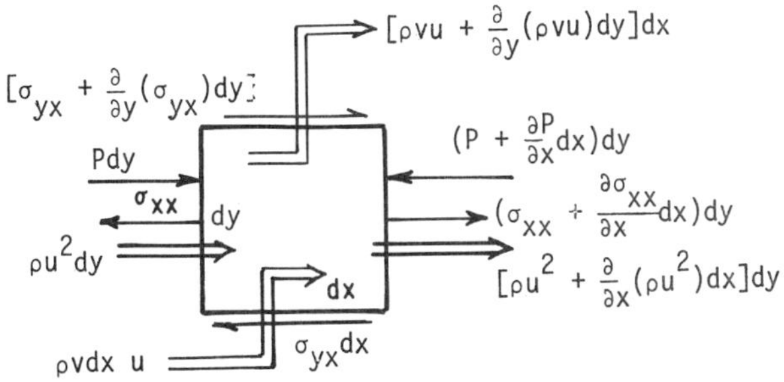

Rate of change of momentumm in x-direction

$$[\rho u^2 + \frac{\partial}{\partial x}(\rho u^2)dx]dy + [\rho vu + \frac{\partial}{\partial y}(\rho vu)dy]dx - \rho u^2 dy - \rho vudx$$

$$= \frac{\partial}{\partial x}(\rho u^2)dxdy + \frac{\partial}{\partial y}(\rho vu)dydx \qquad (1)$$

$$(\Sigma F_x) = -\frac{\partial P}{\partial x}dxdy + \frac{\partial}{\partial x}(\sigma_{xx})dxdy + \frac{\partial}{\partial y}(\sigma_{yx})dydx + \rho f_x \qquad (2)$$

Equating (1) and (2)

$$\frac{\partial}{\partial x}(\rho u^2) + \frac{\partial}{\partial y}(\rho vu) = -\frac{\partial P}{\partial x} + \frac{\partial}{\partial x}(\sigma_{xx}) + \frac{\partial}{\partial y}(\sigma_{yx}) + \rho f_x \qquad (3)$$

with $\sigma_{xx} = 2\mu \frac{\partial u}{\partial x}$, $\sigma_{yx} = \mu(\frac{\partial u}{\partial y} + \frac{\partial v}{\partial x})$, Eq. (3) becomes

$$2\rho u \frac{\partial u}{\partial x} + \rho(v\frac{\partial u}{\partial y} + u\frac{\partial v}{\partial y}) = -\frac{\partial P}{\partial x} + 2\mu\frac{\partial^2 u}{\partial x^2} + \mu\frac{\partial}{\partial y}(\frac{\partial u}{\partial y} + \frac{\partial v}{\partial x}) + \rho f_x$$

$$\rho u \frac{\partial u}{\partial x} + \rho v\frac{\partial u}{\partial y} + \rho u(\frac{\partial u}{\partial x} + \frac{\partial v}{\partial y}) = -\frac{\partial P}{\partial x} + \mu\frac{\partial^2 u}{\partial x^2} + \mu\frac{\partial}{\partial x}(\frac{\partial u}{\partial x} + \frac{\partial v}{\partial y})$$

$$+ \mu\frac{\partial^2 u}{\partial y^2} + \rho f_x \qquad (4)$$

Using the continuity equation and noting that $\frac{\partial^2 u}{\partial x^2} << \frac{\partial^2 u}{\partial y^2}$, and neglecting f_x, Eq. (4) becomes

$$u\frac{\partial u}{\partial x} + v\frac{\partial u}{\partial y} = -\frac{1}{\rho}\frac{\partial P}{\partial x} + \nu(\frac{\partial^2 u}{\partial y^2})$$

Energy equation: Conservation of energy

$$E_{in} = E_{out} = 0$$

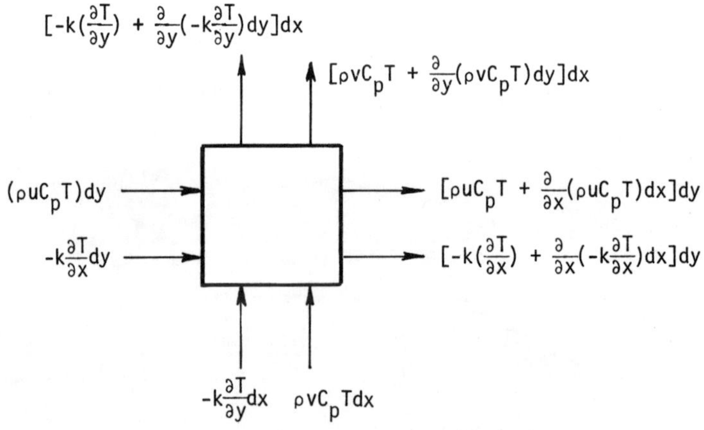

$$- \frac{\partial}{\partial x} (-k \frac{\partial T}{\partial x}) dxdy - \frac{\partial}{\partial x} (\rho u c_p T) dxdy - \frac{\partial}{\partial y} (-k \frac{\partial T}{\partial y}) dydx$$

$$- \frac{\partial}{\partial y} (\rho v c_p T) dydx = 0$$

$$k \frac{\partial^2 T}{\partial x^2} - \frac{\partial}{\partial x} (\rho u c_p T) + k \frac{\partial^2 T}{\partial y^2} - \frac{\partial}{\partial y} (\rho v c_p T) = 0$$

$$k (\frac{\partial^2 T}{\partial x^2} + \frac{\partial^2 T}{\partial y^2}) - \rho c_p [\frac{\partial}{\partial x} (uT) + \frac{\partial}{\partial y} (vT)] = 0$$

$$k (\frac{\partial^2 T}{\partial x^2} + \frac{\partial^2 T}{\partial y^2}) - \rho c_p [u \frac{\partial T}{\partial x} v + \frac{\partial T}{\partial y} + T (\frac{\partial u}{\partial x} + \frac{\partial v}{\partial y})] = 0$$

using the continuity equation, definition of Prandtl number and noting that $\frac{\partial^2 T}{\partial x^2} << \frac{\partial^2 T}{\partial y^2}$ we get

$$u \frac{\partial T}{\partial x} + v \frac{\partial T}{\partial y} = \frac{\nu}{Pr} (\frac{\partial^2 T}{\partial y^2})$$

3.4 $u(\partial p/\partial x)$ is of order $u_e(\rho u_e^2/x)$ while $\partial/\partial y[k(\partial T/\partial y)]$ is of order $k(\Delta T/\delta^2)$. In a gas flow k is nearly proportional to the absolute temperature so that $\partial k/\partial y(\partial T/\partial y)$ is of no larger order than $k(\partial^2 T/\partial y^2)$. Now δ^2 is of order $\nu x/u_e$ so the ratio of $u(\partial p/\partial x)$ to the heat-flux term is of order

$$(\frac{\rho u_e^3}{x})/(\frac{\rho k}{\mu} \frac{u_e \Delta T}{x}) = \frac{u_e^2 Pr}{c_p \Delta T} \quad \text{since} \quad Pr \equiv \frac{\mu c_p}{k}\text{, and the result follows.}$$

3.5 a. With the Mangler transformation given by Eqs. (P3.1),

$\bar{\psi}(\bar{x},\bar{y}) = (1/L)^K \psi(x,y)$ and with the chain rule,

$$u = \frac{1}{r^K} \frac{\partial \psi}{\partial y} = (\frac{L}{r})^K \frac{\partial \bar{\psi}}{\partial \bar{y}} \frac{\partial \bar{y}}{\partial y} = (\frac{L}{r})^K (\frac{r}{L})^K \frac{\partial \bar{\psi}}{\partial \bar{y}} = \bar{u}$$

$$v = - \frac{1}{r^K} \frac{\partial \psi}{\partial x} = - (\frac{L}{r})^K \{\frac{\partial \bar{\psi}}{\partial \bar{x}} \frac{\partial \bar{x}}{\partial x} + \frac{\partial \bar{\psi}}{\partial \bar{y}} \frac{\partial \bar{y}}{\partial x}\} = (\frac{L}{r})^K \{\bar{v}(\frac{r_0}{L})^{2K} - \bar{u} \frac{\partial \bar{y}}{\partial x}\}$$

Here $\bar{u} = \frac{\partial \bar{\psi}}{\partial \bar{y}}$, $\quad \bar{v} = - \frac{\partial \bar{\psi}}{\partial \bar{x}}$

b. With the chain rule,

$$\frac{\partial r^K u}{\partial x} = u \frac{\partial r^K}{\partial x} + r^K \{\frac{\partial u}{\partial \bar{x}} \frac{\partial \bar{x}}{\partial x} + \frac{\partial u}{\partial \bar{y}} \frac{\partial \bar{y}}{\partial x}\}$$

$$= \bar{u} \frac{\partial r^K}{\partial x} + r^K \{(\frac{r_0}{L})^{2K} \frac{\partial \bar{u}}{\partial \bar{x}} + \frac{\partial \bar{u}}{\partial \bar{y}} \frac{\partial \bar{y}}{\partial \bar{x}} \}$$

$$\frac{\partial r^K v}{\partial y} = \frac{\partial}{\partial y}\{L^K \bar{v}\,(\frac{r_0}{L})^{2K} - L^K \bar{u}\,\frac{\partial \bar{y}}{\partial x}\}$$

$$= L^K\{(\frac{r_0}{L})^{2K}\frac{\partial \bar{v}}{\partial \bar{y}}\frac{\partial \bar{y}}{\partial y} - \frac{\partial \bar{u}}{\partial \bar{y}}\frac{\partial \bar{y}}{\partial y}\frac{\partial \bar{y}}{\partial x} - \bar{u}\frac{\partial}{\partial x}(\frac{\partial \bar{y}}{\partial y})\}$$

$$= r^K (\frac{r_0}{L})^{2K}\frac{\partial \bar{v}}{\partial \bar{y}} - r^K\frac{\partial \bar{u}}{\partial \bar{y}}\frac{\partial \bar{y}}{\partial x} - \bar{u}\frac{\partial r^K}{\partial x}$$

The continuity equation expressed in the (x,y) coordinate system can be written as

$$\frac{\partial r^K u}{\partial x} + \frac{\partial r^K v}{\partial y} = 0 = r^K (\frac{r_0}{L})^{2K}(\frac{\partial \bar{u}}{\partial \bar{x}} + \frac{\partial \bar{v}}{\partial \bar{y}}) = \frac{\partial \bar{u}}{\partial \bar{x}} + \frac{\partial \bar{v}}{\partial \bar{y}} = 0$$

Similarly,

$$u\frac{\partial u}{\partial x} = \bar{u}\{\frac{\partial \bar{u}}{\partial \bar{x}}\frac{\partial \bar{x}}{\partial x} + \frac{\partial \bar{u}}{\partial \bar{y}}\frac{\partial \bar{y}}{\partial x}\} = \bar{u}\{(\frac{r_0}{L})^{2K}\frac{\partial \bar{u}}{\partial \bar{x}} + \frac{\partial \bar{u}}{\partial \bar{y}}\frac{\partial \bar{y}}{\partial x}\}$$

$$v\frac{\partial u}{\partial y} = (\frac{L}{r})^K\{\bar{v}\,(\frac{r_0}{L})^{2K} - \bar{u}\frac{\partial \bar{y}}{\partial x}\}\frac{\partial \bar{u}}{\partial \bar{y}}(\frac{r}{L})^K$$

$$\frac{dp}{dx} = \frac{dp}{d\bar{x}}\frac{d\bar{x}}{dx} = \frac{dp}{d\bar{x}}(\frac{r_0}{L})^{2K}$$

$$\nu\frac{1}{r}\frac{\partial}{\partial y}(r\frac{\partial u}{\partial y}) = \nu\frac{1}{r}\frac{\partial}{\partial \bar{y}}\{r\frac{\partial \bar{u}}{\partial \bar{y}}(\frac{r}{L})\}\frac{r}{L} = \nu(\frac{r_0}{L})^2\frac{\partial}{\partial \bar{y}}\{(\frac{r}{r_0})^2\frac{\partial \bar{u}}{\partial \bar{y}}\}$$

Noting the definition of $r = r_0 + y\cos\alpha$, and denoting $t = y\cos\alpha/r_0$

$$(\frac{r}{r_0})^2 = (1 + \frac{y\cos\alpha}{r_0})^2 = (1+t)^2$$

To express t as a function of $\bar{y}$, we write the relation between $\bar{y}$ and y from $d\bar{y} = r/L\,dy = [(r_0 + y\cos\alpha)/L]dy$, $\bar{y} = (r_0/L)y + 1/2L\cos\alpha y^2$ and solve for y with the definition of t given by

$$t = \frac{y\cos\alpha}{r_0} = -1 + \sqrt{1 + \frac{2L\cos\alpha}{r_0^2}\bar{y}}$$

Substituting the above expressions for $u(\partial u/\partial x)$ etc. into the momentum equation yields

$$\bar{u}\frac{\partial \bar{u}}{\partial \bar{x}} + \bar{v}\frac{\partial \bar{u}}{\partial \bar{y}} = -\frac{1}{\rho}\frac{\partial p}{\partial \bar{x}} + \nu\frac{\partial}{\partial \bar{y}}\{(1+t)^{2K}\frac{\partial \bar{u}}{\partial \bar{y}}\}$$

Similarly, the energy equation can be written as

$$\bar{u}\frac{\partial T}{\partial \bar{x}} + \bar{v}\frac{\partial T}{\partial \bar{y}} = \frac{k}{\rho c_p}\frac{\partial}{\partial \bar{y}}[(1+t)^{2K}\frac{\partial T}{\partial \bar{y}}]$$

3.6 The analysis is exactly as in the text up to the point where

$$\rho' = \rho(\gamma - 1)M^2\frac{u'}{u}$$

then multiply by u' and take time average.

3.7 It was shown in the text that

$$\frac{T'}{T_w - T_e} \sim -\frac{u'}{u_e}$$

or, with $\Delta T = T_e - T_w$, $T' \sim \Delta T\, u'/u_e$, $\overline{T'v'} \sim \Delta T\, \overline{u'v'}/u_e$

$$c_p \frac{\partial(\rho \overline{T'v'})}{\partial y} \sim \frac{c_p \Delta T}{u_e} \frac{\partial(\rho \overline{u'v'})}{\partial y} \sim \frac{c_p \Delta T}{u_e} \frac{\tau_w}{\delta}$$

The order of magnitude of the "compression work" $u\, dp/dx$, which is $\sim u_e\, \rho u_e^2/x$, is negligible if it is small when compared with the heat transfer term, i.e.,

$$u_e \frac{\rho u_e^2}{x} << \frac{c_p \Delta T}{u_e} \frac{\tau_w}{\delta}, \quad u_e^2 << c_p \Delta T \frac{x}{\delta} \frac{\tau_w}{\rho u_e^2}$$

Multiply both sides of the second by

$$\frac{1}{(\gamma - 1)c_p T_e}, \quad \frac{u_e^2}{(\gamma - 1)c_p T_e} \equiv M_e^2 << \frac{\Delta T}{(\gamma - 1)T_e} \frac{x}{\delta} \frac{\tau_w}{\rho u_e^2}$$

The order of magnitude of τ_w/u_e^2, can be estimated from the momentum integral equation, (3.67)

$$\theta \sim \frac{c_f}{2} x, \quad \sim \frac{\tau_w}{\rho u_e^2} x$$

Then, $M_e^2 << \dfrac{\Delta T}{(\gamma - 1)T_e} \dfrac{\theta}{\delta}$

Here δ is the boundary-layer thickness and θ the momentum thickness and their ratio δ/θ is ~ 10. Therefore,

$M_e^2 << \dfrac{\Delta T}{10(\gamma - 1)T_e}$ if $u\, dp/dx$ is negligible in the enthalpy equation.

3.8 Neglecting the minor terms in the v-component equation yields

$$\frac{u^2}{R} \sim \frac{1}{\rho} \frac{\partial p}{\partial y}$$

Set ρ and u equal to ρ_e and u_e as an order of magnitude approximation, then

$$\frac{\partial p}{\partial y} \sim \frac{\rho_e u_e^2}{R} \quad \text{or} \quad p_e - p_w \sim \frac{\delta}{R} \rho_e u_e^2$$

Since $a^2 = \gamma p/\rho$, we have

$$\frac{p_e - p_w}{p} \sim \gamma M_e^2 \frac{\delta}{R}$$

3.9 Consider the y-component momentum equation for a boundary layer with convection and viscous diffusion terms neglected,

$$\frac{\partial p}{\partial y} = \frac{\partial \rho\overline{v'^2}}{\partial y}$$

If $\rho\overline{v'^2}$ is of the same order of magnitude as $-\rho\overline{u'v'}$, it is of the same order as τ_w. Then from the above equation,

$$\frac{\Delta p}{p_e} \sim \frac{\tau_w}{p_e} \quad \text{and} \quad c_f \equiv \frac{\tau_w}{1/2\ \rho_e u_e^2} = \frac{\tau_w}{1/2\ \gamma p_e M_e^2}$$

$$\text{so} \quad \frac{\Delta p}{p_e} \sim \gamma M_e^2 \frac{c_f}{2}$$

3.10 If c_f is negligible, (3.68) becomes $\frac{d\theta}{dx} = -\frac{\theta}{u_e}\frac{du_e}{dx}$ (H + 2) and integration with the assumption that H is constant, yields $\ln\theta = -(H + 2)$ $\ln u_e$ + constant, so that $\frac{\theta}{\theta_o} = \left(\frac{u_e}{u_{e,o}}\right)^{-(H+2)}$ where subscript o denotes initial conditions.

3.11 For an incompressible zero pressure-gradient flow, $\frac{du_e}{dx} = 0$ and Eq. (3.78b) reduces to $\frac{d\theta_T}{dx} = St$

3.12 Multiply Eq. (3.10b) by u and integrate it across the layer,

$$\int_0^\infty \rho u^2 \frac{\partial u}{\partial x} dy + \int_0^\infty \rho uv \frac{\partial u}{\partial y} dy = \int_0^\infty \rho u u_e \frac{du_e}{dx} dy + \int_0^\infty \mu u \frac{\partial^2 u}{\partial y^2} dy \qquad (1)$$

and then evaluate each term

$$\int_0^\infty \rho u^2 \frac{\partial u}{\partial x} dy = \frac{1}{2}\int_0^\infty \rho u \frac{\partial u^2}{\partial x} dy = \frac{1}{2}\int_0^\infty \rho\left(\frac{\partial u^3}{\partial x} - u^2\frac{\partial u}{\partial x}\right)dy \qquad (2a)$$

$$\int_0^\infty \rho uv \frac{\partial u}{\partial y} dy = \frac{1}{2}\int_0^\infty \rho v \frac{\partial u^2}{\partial y} dy = \frac{1}{2}\left\{\rho v u^2 \Big|_0^\infty - \int_0^\infty u^2 \frac{\partial v}{\partial y} dy\right\}$$

$$= \frac{1}{2}\left\{- u_e^2 \int_0^\infty \frac{\partial u}{\partial x} dy - \int_0^\infty u^2 \frac{\partial v}{\partial y} dy\right\} \qquad (2b)$$

With $v_e = -\int_0^\infty \frac{\partial u}{\partial x}$ dy from the continuity equation,

$$\int_0^\infty \rho u u_e \frac{du_e}{dx} dy = \frac{1}{2}\int_0^\infty \rho u \frac{du_e^2}{dx} dy \qquad (2c)$$

$$\int_0^\infty \mu u \frac{\partial^2 u}{\partial y^2} dy = \mu u \frac{\partial u}{\partial y}\Big|_0^\infty - \mu\int_0^\infty \left(\frac{\partial u}{\partial y}\right)^2 dy \qquad (2d)$$

Substituting (2a) - (2d) into (1), we have

$$\frac{1}{2}\int_0^\infty \left(\frac{\partial u^3}{\partial x} - u^2 \frac{\partial u}{\partial x}\right) dy - \frac{1}{2}\int_0^\infty u_e^2 \frac{\partial u}{\partial x}\, dy - \frac{1}{2}\int_0^\infty u^2 \frac{\partial v}{\partial y}\, dy$$

$$- \frac{1}{2}\int_0^\infty \rho u \frac{du_e^2}{dx}\, dy = -\mu \int_0^\infty \left(\frac{\partial u}{\partial y}\right)^2 dy$$

After rearranging and using the continuity equation, $\partial u/\partial x + \partial v/\partial y = 0$, we obtain

$$\frac{1}{2}\int_0^\infty \rho \frac{\partial}{\partial x}(u^3 - uu_e^2)\, dy = -\mu \int_0^\infty \left(\frac{\partial u}{\partial y}\right)^2 dy$$

which can be written in the same form as (P3.7) by using the definition of the kinetic-energy thickness δ^* given by (P3.8).

Chapter 4

Uncoupled Laminar Boundary Layers

4.1 Using the Falkner-Skan transformation given by (4.19b) and (4.19c), and the definition of dimensionless temperature g given by (4.24), it follows from the chain rule that

$$u = \left(\frac{\partial \psi}{\partial y}\right)_x = \frac{\partial \psi}{\partial \eta}\frac{\partial \eta}{\partial y} = u_e f'$$

$$v = -\left(\frac{\partial \psi}{\partial x}\right)_y = -\left\{ \left(\frac{\partial \psi}{\partial x}\right)_\eta + \frac{\partial \psi}{\partial \eta}\frac{\partial \eta}{\partial x} \right\}$$

$$= -(u_e \nu x)^{1/2} \left\{ \frac{\partial f}{\partial x} + \frac{1}{2}\left(\frac{1}{u_e}\frac{du_e}{dx} + \frac{1}{x}\right) f + f' \frac{\partial \eta}{\partial x} \right\}$$

$$u\frac{\partial u}{\partial x} = u_e f' \left\{ \left(\frac{\partial u}{\partial x}\right)_\eta + \frac{\partial u}{\partial \eta}\frac{\partial \eta}{\partial x} \right\}$$

$$= u_e f' \left\{ \frac{du_e}{dx} f' + u_e \frac{\partial f'}{\partial x} + u_e f'' \frac{\partial \eta}{\partial x} \right\}$$

$$v\frac{\partial u}{\partial y} = v \frac{\partial u}{\partial \eta}\frac{\partial \eta}{\partial y} = -u_e^2 \left\{ f'' \frac{\partial f}{\partial x} + \frac{1}{2x}\left(\frac{x}{u_e}\frac{du_e}{dx} + 1\right) f f'' + f' f'' \frac{\partial \eta}{\partial x} \right\}$$

$$\nu \frac{\partial^2 u}{\partial y^2} = \frac{u_e^2}{x} f'''$$

$$u\left(\frac{\partial T}{\partial x}\right)_y = u_e f' \left\{ (1 - g)\frac{d(T_w - T_e)}{dx} - (T_w - T_e)\frac{\partial g}{\partial x} \right\}$$

$$- u_e (T_w - T_e) f' g' \frac{\partial \eta}{\partial x}$$

$$v\frac{\partial T}{\partial y} = u_e (T_w - T_e)\left\{ g' \frac{\partial f}{\partial x} + \frac{1}{2x}\left(\frac{x}{u_e}\frac{du_e}{dx} + 1\right) f g' + f' g' \frac{\partial \eta}{\partial x} \right\}$$

$$\frac{\nu}{Pr}\frac{\partial^2 T}{\partial y^2} = -\frac{(T_w - T_e)}{Pr}\left(\frac{u_e}{\nu x}\right) g''$$

Substituting the above relations into (3.10b) and (3.11) and, after some arrangement, we obtain the expressions given by (4.21) and

(4.28) with m and n given by (4.23) and (4.29a). For similar flows, f and g are functions of η only and, as a result, (4.21) and (4.28) reduce to (4.31) and (4.32) subject to the boundary conditions given by

$$\eta = 0: \quad f = f' = 0, \quad g = 0; \qquad \eta \to \infty: \quad f' = 1, \quad g = 1$$

4.2 a. In terms of the stream function, ψ, the momentum equation and its normal boundary condition at the wall can be written as

$$\frac{\partial\psi}{\partial y}\frac{\partial^2\psi}{\partial x\partial y} - \frac{\partial\psi}{\partial x}\frac{\partial^2\psi}{\partial y^2} = u_e\frac{du_e}{dx} + \nu\frac{\partial^3\psi}{\partial y^3}, \quad v_w = -\left(\frac{\partial\psi}{\partial x}\right)_w \tag{1}$$

Introducing a linear transformation by

$$x = A^{\alpha_1}\bar{x}, \quad y = A^{\alpha_2}\bar{y}, \quad \psi = A^{\alpha_3}\bar{\psi}, \quad u_e = A^{\alpha_4}\bar{u}_e, \quad v_w = A^{\alpha_5}\bar{v}_w$$

we can write the two equations in (1) as

$$A^{2\alpha_3-\alpha_1-2\alpha_2}\left(\frac{\partial\bar{\psi}}{\partial\bar{y}}\frac{\partial^2\bar{\psi}}{\partial\bar{x}\partial\bar{y}} - \frac{\partial\bar{\psi}}{\partial\bar{x}}\frac{\partial^2\bar{\psi}}{\partial\bar{y}}\right) = A^{2\alpha_4-\alpha_1}\bar{u}_e\frac{d\bar{u}_e}{d\bar{x}} + \nu A^{\alpha_3-3\alpha_2}\frac{\partial^3\bar{\psi}}{\partial\bar{y}^3}$$

$$A^{\alpha_5}\bar{v}_w = -A^{\alpha_3-\alpha_1}\frac{\partial\bar{\psi}}{\partial\bar{x}}.$$

The invariance under the transformation requires that

$$2\alpha_3 - \alpha_1 - 2\alpha_2 = 2\alpha_4 - \alpha_1 = \alpha_3 - 3\alpha_2, \qquad \alpha_5 = \alpha_3 - \alpha_1$$ from

which, $\frac{\alpha_3}{\alpha_1} = 1 - \alpha, \qquad \frac{\alpha_4}{\alpha_1} = 1 - 2\alpha, \qquad \frac{\alpha_5}{\alpha_1} = \frac{\alpha_3}{\alpha_1} - 1 = -\alpha$

where $\alpha = \frac{\alpha_2}{\alpha_1}$

Therefore, for similarity $\frac{v_w}{x^{-\alpha}} = \frac{\bar{v}_w}{\bar{x}^{-\alpha}}$

Since v_w is a function of x only, $v_w = c_w x^{-\alpha}$

For a similar flow, $u_e = cx^m$ where $\alpha = (1 - m)/2$. Therefore

$$v_w = c_w x^{(m-1)/2}$$

b. The dimensionless stream function at the wall given by (4.30a) is a constant for a similar flow and $u_e(x) \sim x^m$ so that

$$\int_0^x v_w dx \sim x^{(m+1)/2}$$

Taking the derivative of both sides with respect to x, $v_w \sim x^{(m-1)/2}$ as shown in Part a.

c. For a laminar flat-plate flow, $m = 0$ and $\alpha = -1/2$ and, from Part b., $f_w = -(2v_w x)/(u_e \nu x)^{1/2} = -2(v_w/u_e)\sqrt{R_x}$.

4.3 Let $t = A^{\alpha_1}\bar{t}$, $y = A^{\alpha_2}\bar{y}$, $v = A^{\alpha_3}\bar{v}$, then under the linear transformation, the governing equation can be written as

$$A^{\alpha_3-2\alpha_1-2\alpha_2}\left(\frac{\partial}{\partial \bar{t}} - A^{\alpha_1+\alpha_2}\, i\bar{y}\right)^2 \frac{\partial^2 \bar{v}}{\partial \bar{y}^2} = A^{\alpha_3} N^2 \bar{v}$$

The invariance under the transformation requires that

$$\alpha_2 + \alpha_1 = 0, \quad \alpha_3 - 2\alpha_1 - 2\alpha_2 = \alpha_3$$

from which $\alpha = \frac{\alpha_2}{\alpha_1} = -1$ and the similarity parameter is $\eta = \frac{y}{t^\alpha} = ty$ and v can be written as $v = t^n F(\eta)$

$$\frac{\partial^2 v}{\partial y^2} = t^n \frac{\partial}{\partial \eta}\left(\frac{\partial F}{\partial \eta}\frac{\partial \eta}{\partial y}\right)\frac{\partial \eta}{\partial y} = t^{n+2}F''$$

$$\left(\frac{\partial}{\partial t} - iy\right)^2 \frac{\partial^2 v}{\partial y^2} = \left(\frac{\partial}{\partial t} - iy\right)\left\{\left(\frac{\partial}{\partial t} - iy\right)(t^{n+2}F'')\right\}$$

$$= \left(\frac{\partial}{\partial t} - iy\right)\{(n+2)t^{n+1}F'' + yt^{n+2}F''' - iyt^{n+2}F''\}$$

$$= t^n\{\eta^2 F'''' + 2\eta F'''(n + 2 - i\eta) + F''[(n+2)(n+1) - 2i\eta(n+2) - \eta^2]\}$$

so: $\eta^2 F'''' + 2\eta F'''(n + 2 - i\eta) + F''[(n+2)(n+1) - 2i\eta(n+2) - \eta^2] = N^2 F$

4.4 In the new coordinate system (x,ψ) the terms of (3.10b) and (3.11) with $u = \frac{\partial \psi}{\partial y}$, $v = -\frac{\partial \psi}{\partial x}$, can be expressed as

$$u\left(\frac{\partial u}{\partial x}\right)_y = u\left[\left(\frac{\partial u}{\partial x}\right)_\psi + \frac{\partial u}{\partial \psi}\frac{\partial \psi}{\partial x}\right] = u\frac{\partial u}{\partial x} - uv\frac{\partial u}{\partial \psi}$$

$$v\frac{\partial u}{\partial y} = v\frac{\partial u}{\partial \psi}\frac{\partial \psi}{\partial y} = uv\frac{\partial u}{\partial \psi}$$

$$\nu\frac{\partial^2 u}{\partial y^2} = \nu\frac{\partial}{\partial \psi}\left(\frac{\partial u}{\partial \psi}\frac{\partial \psi}{\partial y}\right)\frac{\partial \psi}{\partial y} = \nu u\frac{\partial}{\partial \psi}\left(u\frac{\partial u}{\partial \psi}\right)$$

$$u\frac{\partial T}{\partial x} = u\left[\left(\frac{\partial T}{\partial x}\right)_\psi + \frac{\partial T}{\partial \psi}\frac{\partial \psi}{\partial x}\right] = u\frac{\partial T}{\partial x} - uv\frac{\partial T}{\partial \psi}$$

$$v\frac{\partial T}{\partial y} = uv\frac{\partial T}{\partial \psi}, \qquad \frac{\nu}{Pr}\frac{\partial^2 T}{\partial y^2} = \frac{\nu}{Pr}u\frac{\partial}{\partial \psi}\left(u\frac{\partial T}{\partial \psi}\right)$$

Substituting the above relations into (3.10b) and (3.11) we obtain (P4.1) and (P4.2).

4.5 a. Integrating (4.31) with respect to η from 0 to η_e, we get

$$\int_0^{\eta_e} f'''d\eta + \frac{m+1}{2}\int_0^{\eta_e} ff''d\eta + m\int_0^{\eta_e}(1 - f'^2)d\eta = 0 \qquad (1)$$

Noting that $\int_0^{\eta_e} f'''d\eta = f''\Big|_0^{\eta_e} = -f''_w$

$$m\int_0^{\eta_e}(1 - f'^2)d\eta = m\int_0^{\eta_e}[1 - f' + f'(1 - f')]d\eta = m(\delta_1^* + \theta_1)$$

Eq. (1) can be written in the form given by (P4.4).

b. Integrating by parts,

$$\int_0^{\eta_e} ff''d\eta = ff'\Big|_0^{\eta_e} - \int_0^{\eta_e}[f' - f'(1 - f')]d\eta$$

$$= f_e - [f_e - f_w - \theta_1] = f_w + \theta_1$$

where f_w is the value of f at the wall and is zero when there is no mass transfer. Thus, $\int_0^{\eta_e} ff''d\eta = \theta_1$ and (P4.4) reduces to that given by (P4.5).

4.6 For constant wall temperature, (4.32) reduces to (4.41). Integrating the first term of this equation with respect to η from 0 to η_e and using the relations,

$$\int_0^{\eta_e} g''\,d\eta = (g'_e - g'_w) = -g'_w$$

and $$Pr\,\frac{m+1}{2}\int_0^{\eta_e} fg'd\eta = Pr\,\frac{m+1}{2}\left[(fg)_0^{\eta_e} - \int_0^{\eta_e} gf'd\eta\right]$$

$$= Pr\,\frac{m+1}{2}\{f_e - f_e + f_w + \theta_T\}$$

we get (P4.7) with θ_{T_1} as defined in problem (P4.6).

4.7 Integrating (4.31) and (4.32) wrt η from 0 to η_e, we get

$$-f''_w + \frac{m+1}{2}\int_0^{\eta_e} ff''d\eta + m(\delta_1^* + \theta_1) = 0 \qquad (1)$$

$$\frac{-g'_w}{Pr} + \frac{m+1}{2}\int_0^{\eta_e} fg'd\eta + n\int_0^{\eta_e} f'_1(1 - g)d\eta = 0 \qquad (2)$$

Integration by parts yields

$$\frac{m+1}{2}\int_0^{\eta_e} ff''d\eta = \frac{m+1}{2}\left\{ff'\Big|_0^{\eta_e} - \int_0^{\eta_e}[f' - f'(1 - f')]d\eta\right\}$$

$$= \frac{m+1}{2}(f_w + \theta_1) \qquad (3)$$

$$\frac{m+1}{2}\int_0^{\eta_e} fg'd\eta = \frac{m+1}{2}\{fg\Big|_0^{\eta_e} - \int_0^{\eta_e}[f' - f'(1-g)]d\eta\}$$

$$= \frac{m+1}{2}\{f_e - f_e + f_w + \int_0^{\eta_e} f'(1-g)d\eta\} = \frac{m+1}{2}(f_w + \theta_T) \quad (4)$$

Using the definitions of θ_1 and θ_T and substituting (3) into (1) and (4) into (2) we get the relations given by (P4.8) and (P4.9).

4.8 a. With $\rho = 1.009$ kg/m^3, $\nu = 2.0 \times 10^{-5}$m^2/s, $c_p = 1009$ J/kg K and and Pr = 0.72 for air @ $T_f = \frac{1}{2}(20 + 120) = 70°C$, and for a zero-pressure gradient flow with a uniform wall temperature

$$c_f = 0.664/\sqrt{R_x}, \quad g_w' = Nu_x/\sqrt{R_x} = 0.2957$$

With $R_x = \frac{u_e x}{\nu} = \frac{10.0 \times 1.5}{2 \times 10^{-5}} = 7.5 \times 10^5$,

$$c_f = 0.664/\sqrt{7.5 \times 10^5} = 0.767 \times 10^{-3}$$

$$Nu_x = 0.2957\sqrt{R_x} = 2.56 \times 10^2$$

$$St_x = \frac{Nu_x}{PrR_x} = \frac{2.56 \times 10^2}{(0.72 \times 7.5 \times 10^5)} = 0.474 \times 10^{-3}$$

b. The total heat flux through the plate is given by

$$Q_w = \int_0^L W\dot{q}_w dx$$

where W is the plate width and $\dot{q}_w$ is the local heat transfer rate given by $\dot{q} = -k(\partial T/\partial y)_w = \rho c_p(T_w - T_e)(u_e/Pr)(g_w'/\sqrt{R_x})$

$$\therefore Q_w = W\rho c_p(T_w - T_e)\frac{u_e}{Pr}g_w'\frac{2x}{\sqrt{R_x}}\Bigg|_{x=0}^{L}$$

$$= \frac{0.5 \times 1.009 \times 1009.0 \times (120 - 20) \times 10}{0.72} \times 0.2957$$

$$\times \frac{2 \times 1.5}{\sqrt{7.5 \times 10^5}} = 724 \text{ W}$$

4.9 With $\rho = 1258$ kg/m^3, $\nu = 0.5 \times 10^{-3}$m^2/s, $c_p = 2443.0$ J/kg K and Pr = 5380 for glycerine @ $T_f = \frac{1}{2}(T_w + T_e) = 30°C$, and since Pr << 1, $Nu_x = 0.339\,Pr^{1/3}R_x^{1/2}$ and the local heat transfer rate is then (from the plate)

$$\dot{q}_w = St\rho c_p(T_w - T_e)u_e = \frac{Nu_x}{PrR_x}\rho c_p(T_w - T_e)u_e$$

$$= 0.339Pr^{-2/3}R_x^{-1/2}\rho c_p(T_w - T_e)u_e$$

The total heat transfer to the plate is

$$q_w = -\int_0^L \dot{q}_w dx = 0.339\ Pr^{-2/3} \rho c_p (T_e - T_w) u_e \frac{2L}{\sqrt{R_L}} = 0.339(5380)^{-2/3}$$

$$\times\ 1258 \times 2443 \times (40 - 20) \times 3 \times \frac{2 \times 4}{\sqrt{(3 \times 4)/(0.5 \times 10^{-3})}}$$

$$= 1.0513 \times 10^4\ W/m$$

4.10 a. The Nusselt number based on the cylinder radius is

$$Nu_{r_0} = \frac{r_0 q_w}{k(T_w - T_e)} = -\frac{r_0(\partial T/\partial y)_w}{T_w - T_e}$$

When the wall temperature is specified, we define the dimensionless temperature, g, by (4.24) and the similarity parameter η by (4.19b) and write

$$(\frac{\partial T}{\partial y})_w = (T_e - T_w) \frac{\partial g}{\partial \eta} \frac{\partial \eta}{\partial y} = (T_e - T_w) \sqrt{\frac{u_e}{\nu x}}\, g_w'$$

so that the expression for the Nusselt number, with $u_e = (2u_\infty x/r_0)$ and $R = (u_\infty r_0/\nu)$, becomes

$$Nu_{r_0} = -\frac{r_0}{T_w - T_e} (T_e - T_w) \sqrt{\frac{u_e}{\nu x}}\, g_w' = r_0 \sqrt{\frac{2u_\infty x}{r_0} \frac{1}{\nu x}}\, g_w' = \sqrt{2}\sqrt{R}\, g_w'$$

b. With δ^* and θ_T representing the displacement and enthalpy thicknesses,

$$\delta^* = x\delta_1^*/\sqrt{R_x}, \quad \theta_T = x\theta_{T_1}/\sqrt{R_x}$$

where $\delta_1^* = \int_0^\infty (1 - f')d\eta, \qquad \theta_{T_1} = \int_0^\infty f'(1 - g)d\eta$

At a stagnation point, $m = 1$ and $\delta_1^* = 0.64791$, $\theta_{T_1} = g_w'/Pr = 0.5017/0.72 = 0.697$, so that with $u_e = 2u_\infty x/r_0$,

$$\delta^* = \frac{x}{\sqrt{2u_\infty x/r_0\ x/\nu}} \delta_1^* = r_0/\sqrt{2u_\infty r_0/\nu}\ \delta_1^*$$

$$= 0.01/\sqrt{\frac{2 \times 1.0 \times 0.01}{1.5 \times 10^{-5}}} \times 0.64791 = 0.177mm$$

$$\theta_T = \frac{r_0}{\sqrt{2u_\infty r_0/\nu}} \theta_{T_1} = 0.01/\sqrt{\frac{2 \times 1.0 \times 0.01}{1.5 \times 10^{-5}}} \times 0.697 = 0.191mm$$

4.11 The equivalent flow in two dimensions is given by the Mangler transformation (no transverse curvature effect), namely by

$$d\bar{x} = (\frac{r_0}{L})^2 ds, \quad \bar{y} = (\frac{r_0}{L})y \qquad (1)$$

where r_0 and s are related by the cone half-angle β_c, that is, $r_0 = s \cdot \sin\beta_c$. To find the required relation, we determine the equivalent pressure gradient parameter m for two-dimensional flow. Noting that the exponent 2.2 is equal to m and that:

$$2.2 = \frac{s}{u_e}\frac{du_e}{ds} = \frac{s}{\bar{x}}\frac{\bar{x}}{u_e}\frac{du_e}{d\bar{x}}\frac{d\bar{x}}{ds} = \frac{s}{\bar{x}}\frac{d\bar{x}}{ds}\, m,$$

$$\text{and } \bar{x} = \int_0^s \frac{\sin^2\beta_c}{L^2} s^2 ds = \frac{\sin^2\beta_c}{L^2}\frac{s^3}{3}$$

$$\text{so that } 2.2 = \frac{s}{(\sin\beta_c/L)^2(s^3/3)} s^2 (\sin\beta_c/L)^2 m = 3m \text{ or } m = \frac{2.2}{3} = 0.733$$

For the equivalent two-dimensional flow we write δ_1^* as

$$\delta_1^* = \delta_2^* \sqrt{\frac{u_e}{\nu\bar{x}}} \tag{2}$$

where δ_1^* depends on m only and is obtained from Table 4.1 by interpolation: $\delta_1^* = 0.74$. To obtain the displacement thickness for the cone, we transform this relationship back to the cone surface by using (1) and noting that $\delta_2^* = (r_0/L)\,\delta_c^*$, we obtain

$$\delta_1^* = \delta_c^* \sqrt{\frac{3u_e}{\nu s}} \quad \text{or} \quad \frac{0.427s}{\sqrt{u_e s/\nu}} = \delta_c^*$$

4.12 To derive (4.53), we start with (4.45) which is valid for all Pr and m. For Pr >> 1, take $f' = f_w''\eta$ so that $f = \eta^2/2\, f_w''$. From the definition of Nusselt number

$$Nu = \left(\frac{\partial g}{\partial \eta}\right)_w \sqrt{R_x} = R_x^{1/2} \left\{ \int_0^{\eta_e} \exp\left[-Pr\,(m+1)/2 \int_0^{\zeta} f(z)dz\right] d\zeta \right\}^{-1}$$

and from the expression for f, it follows that

$$\int_0^{\eta_e} \exp\left[-Pr\,\frac{m+1}{2}\int_0^{\zeta} f(z)dz\right]d\zeta = \int_0^{\infty} \exp\left[-Pr\,\frac{m+1}{12} f_w''\,\zeta^3\right]d\zeta \tag{1}$$

Let $t = Pr\,\frac{m+1}{12} f_w''\,\zeta^3$ so that $\frac{1}{3}\left(Pr\,\frac{m+1}{12} f_w''\right)^{-1/3} t^{-2/3} dt = d\zeta$

The RHS of (1) becomes

$$\int_0^{\infty} \exp\left[-Pr\,\frac{m+1}{12} f_w''\zeta^3\right]d\zeta = \frac{1}{3} Pr^{-1/3}\left(\frac{m+1}{12} f_w''\right)^{-1/3} \int_0^{\infty} e^{-t} t^{(1/3)-1} dt$$

$$= \frac{1}{3}\Gamma(1/3) Pr^{-1/3}\left(\frac{m+1}{12} f_w''\right)^{-1/3}$$

so that with $\Gamma(1/3) = 2.6789$,

$$Nu = R_x^{1/2} Pr^{1/3} \left(\frac{m+1}{12} f_w''\right)^{1/3} \frac{3}{\Gamma(1/3)} = 1.12 \left(\frac{m+1}{12} f_w''\right)^{1/3} R_x^{1/2} Pr^{1/3}$$

4.13 a. Consider the energy equation, (4.10) and its boundary conditions $y = 0$: $g = 0$ for $x < x_0$, $g = 1$ for $x > x_0$ and $y = \delta_t$: $g = 0$ for $x > x_0$. Define a new variable ζ such that

$$\zeta = y/z^{1/3}, \quad z = x - x_0 \quad \text{so that}$$

$$u = \sum_{k=1}^{\infty} \lambda_k y^k = \sum_{k=1}^{\infty} \lambda_k z^{k/3}\zeta^k, \quad v = \sum_{k=2}^{\infty} \mu_k y^k = \sum_{k=2}^{\infty} \mu_k z^{k/3}\zeta^k$$

$$g = \sum_{k=0}^{\infty} z^{k/3} G_k(\zeta)$$

chain rule gives

$$\left(\frac{\partial g}{\partial x}\right)_y = \left(\frac{\partial g}{\partial x}\right)_\zeta - \frac{1}{3}\frac{\zeta}{z} g', \quad \left(\frac{\partial g}{\partial y}\right)_x = z^{-1/3} g', \quad \left(\frac{\partial^2 g}{\partial y^2}\right) = z^{-2/3} g''$$

Substituting the relations into (P4.10) yields

$$\sum_{k=1}^{\infty} \lambda_k z^{k/3}\zeta^k \left[\sum_{n=0}^{\infty} z^{n/3}\left(z\frac{\partial G_n}{\partial x} - \frac{1}{3}\zeta G_n'\right)\right]$$

$$+ z^{2/3}\left(\sum_{k=2}^{\infty} \mu_k z^{k/3}\zeta^k\right)\left(\sum_{n=0}^{\infty} z^{n/3} G_n'\right) = \frac{\nu}{Pr} z^{1/3} \sum_{n=0}^{\infty} z^{n/3} G_n''$$

Now expand the above expression and consider the lowest order of z, i.e, for $k = 1$ and $n = 0$,

$$-\frac{1}{3}\lambda_1 \zeta^2 G_0' = \frac{\nu}{Pr} G_0'' \quad \text{or} \quad G_0'' + \frac{1}{3}\lambda_1 \frac{Pr}{\nu}\zeta^2 G_0' = 0 \qquad \text{(P4.15)}$$

b. Integrate (P4.15) twice with respect to ξ,

$$G_0 = b + a\int_0^\zeta \exp\left\{-\frac{1}{9}\lambda_1 \frac{Pr}{\nu}\zeta^3\right\} d\zeta$$

The integration constants a and b are determined from

$\zeta = 0$, $G_0 = 1$, $\zeta \to \infty$, $G_0 \to 0$

from which $b = 1.0$, $a = -\frac{3}{\Gamma(1/3)}\left(\frac{1}{9}\lambda_1 \frac{Pr}{\nu}\right)^{1/3}$

Hence $$G_0 = 1 - \frac{3}{\Gamma(1/3)}\left(\frac{1}{9}\lambda_1\frac{Pr}{\nu}\right)^{1/3}\int_0^\zeta \exp\left(-\frac{1}{9}\lambda_1\frac{Pr}{\nu}\zeta^3\right)d\zeta \qquad \text{(P4.16)}$$

c. With $u = \sum_{k=1}^{\infty} \lambda_k y^k$, $\frac{\partial u}{\partial y} = \lambda_1 + \sum_{k=2}^{\infty} k\lambda_k y^{k-1}$

At $y = 0$, $\lambda_1 = \left(\frac{\partial u}{\partial y}\right)_w = f_w'' \frac{u_e}{x}\sqrt{\frac{u_e x}{\nu}}$

For a flat-plate laminar flow at $x = x_0$ with $f_w'' = 0.332$

$$\lambda_1 = 0.332 \frac{u_e}{x_0}\sqrt{\frac{u_e x_0}{\nu}}$$

d. $\dot{q}_w = -k\left(\frac{\partial T}{\partial y}\right)_w, \quad \simeq -k(T_w - T_e)\left(\frac{\partial G_o}{\partial \zeta}\right)_w \frac{\partial \zeta}{\partial y}$

$$= \frac{k}{z^{1/3}}(T_w - T_e)\frac{3}{\Gamma(1/3)}\left(\frac{1}{9}\lambda_1 \frac{Pr}{\nu}\right)^{1/3}, \quad \text{from part (a)}$$

$$Nu_{x_0} = \frac{\dot{q}_w}{T_w - T_e}\frac{x_0}{k} = \frac{x_0}{z^{1/3}}\frac{3}{\Gamma(1/3)}\left(\frac{1}{9}\lambda_1 \frac{Pr}{\nu}\right)^{1/3}$$

$$= \frac{3}{\Gamma(1/3)}\left\{\frac{Pr}{\nu} \times \frac{0.332}{9}\frac{u_e}{x_0}\sqrt{\frac{u_e x_0}{\nu}}\right\}^{1/3}\frac{x_0}{(x - x_0)^{1/3}}$$

$$= cR_{x_0}^{1/2}Pr^{1/3}/\left(\frac{x}{x_0} - 1\right)^{1/3}$$

where $c = \frac{3}{\Gamma(1/3)}\left(\frac{0.332}{9}\right)^{1/3}, \quad R_{x_0} = \frac{u_e x_0}{\nu}$

4.14 Write (4.65) as $S^3 + \frac{4}{3}x\frac{dS^3}{dx} = \frac{13}{14}\frac{1}{Pr}$

With $Y = S^3 - \frac{13}{14}\frac{1}{Pr}$ we have $Y + \frac{4}{3}x\frac{dY}{dx} = 0$

so that $Y = cx^{-3/4}$ or $S^3 = \frac{13}{14}Pr^{-1} + cx^{-3/4}$

At $x = x_0$, $S = 0$, $0 = cx_o^{-3/4} + \frac{13}{14}Pr^{-1}$ $\therefore$ $c = -\frac{13}{14}Pr^{-1}x_0^{3/4}$

$\therefore$ $S^3 = \frac{13}{14}Pr^{-1}\left[1 - \left(\frac{x_0}{x}\right)^{3/4}\right]$ or $S = \frac{1}{1.026}Pr^{-1/3}\left[1 - \left(\frac{x_0}{x}\right)^{3/4}\right]^{1/3}$

4.15 a. Integrate the enthalpy integral equation,

$$\frac{d}{dx}\{u_e\theta_T(T_w - T_e)\} = \frac{\dot{q}_w}{\rho c_p} \quad \text{wrt } x \text{ to get}$$

$$u_e\theta_T(T_w - T_e) = \frac{x\dot{q}_w}{\rho c_p} + c \tag{1}$$

at $x = x_0$, $\theta_T = 0$, so $c = -\frac{\dot{q}_w x_0}{\rho c_p}$,

$$\theta_T = \frac{\dot{q}_w(x - x_0)}{\rho c_p u_e(T_w - T_e)} \tag{2}$$

To find θ_T, use the velocity and temperature profiles given by

$$\frac{u}{u_e} = \frac{3}{2}\frac{y}{\delta} - \frac{1}{2}\left(\frac{y}{\delta}\right)^3, \quad \frac{T - T_e}{T_w - T_e} = 1 - \frac{3}{2}\frac{y}{\delta_t} + \frac{1}{2}\left(\frac{y}{\delta_t}\right)^3$$

which, with $S = \delta_t/\delta$, results in

$$\theta_T = \delta\left(\frac{3}{20}S^2 - \frac{3}{280}S^4\right) \tag{3}$$

Since the second term in (3) is small compared with the first term, θ_T can be approximated as

$$\theta_T = \frac{3}{20}\,\delta_t^2/\delta, \quad \text{with } \delta_t \text{ given by} \quad \delta_t = \frac{3k(T_w - T_e)}{\dot{q}_w} \qquad (4)$$

From (2) and (4) $\dfrac{\dot{q}_w^3}{k^3(T_w - T_e)^3} = \dfrac{27}{80 \times 4.64} R_x^{3/2} Pr(1 - \dfrac{x_0}{x})^{-1}/x^3$

$$\therefore \quad Nu_x \equiv \frac{\dot{q}_w x}{k(T_w - T_e)} = 0.4174\, R_x^{1/2} Pr^{1/3} \left(1 - \frac{x_0}{x}\right)^{-1/3}$$

b. From part (a), $T_w - T_e = \frac{x}{k}\dot{q}_w \left\{0.417\, R_x^{1/2} Pr^{1/3} (1 - \frac{x_0}{x})^{-1/3}\right\}^{-1}$

$$= 2.40\, \dot{q}_w \frac{x}{k} R_x^{-1/2} Pr^{-1/3} \left(1 - \frac{x_0}{x}\right)^{1/3}$$

4.16 The total power, P, required to maintain the flow at 80°C is equal to the total heat transfer through the wall, $P = Q_w = \int_{x_0}^{\ell} \dot{q}_w W dx$,

with $\dot{q}_w = (T_w - T_e)\hat{h} = \dfrac{0.332 k Pr^{1/3}\sqrt{R_x}}{x[1 - (x_0/x)^{3/4}]^{1/3}}\,(T_w - T_e)$

$$P = 0.332 w k Pr^{1/3}(T_w - T_e)\sqrt{R_{x_0}} \int_{x_0}^{\ell} \frac{d(x/x_0)}{(x/x_0)^{1/4}[(x/x_0)^{3/4} - 1]^{1/3}}$$

where w is the width of the plate and $R_{x_0} = \dfrac{u_e x_0}{\nu}$

Define $\dfrac{x}{x_0} = x'$, $\quad z^3 = x'^{3/4} - 1$

Then $\displaystyle\int \frac{d(x/x_0)}{(x/x_0)^{1/4}[(x/x_0)^{3/4} - 1]^{1/3}} = \int \frac{4z^2 dz}{z} = 2\left[\left(\frac{x}{x_0}\right)^{3/4} - 1\right]^{2/3}$

$$\therefore \quad P = 0.332 w k Pr^{1/3}(T_w - T_e)\sqrt{R_{x_0}} \left\{2\left[\left(\frac{L}{x_0}\right)^{3/4} - 1.0\right]^{2/3}\right\}$$

With $T_w = 80°C$, $T_e = 20°C$, $T_m = \frac{1}{2}(T_w + T_e) = 50°C$,

$k = 0.028 W/m\,K$, $\nu = 18.2 \times 10^{-6} m^2/s$, $Pr = 0.72$,

$u_e = 10.0 m/s$, $w = 0.4 m$, $L = 1.5 m$, $\quad x_0 = 1.0 m$

$$P = 0.332 \times 0.4 \times 0.028 \times (0.72)^{1/3}(80 - 20)\left(\frac{10.0 \times 1.0}{18.2 \times 10^{-6}}\right)^{1/2}$$

$$\times\, 2.0(1.5^{3/4} - 1.0)^{2/3} = 148.7\ W$$

4.17 $$\dot{q}_w = \sum_{i=1}^{3} \hat{h}_i \Delta T_{w_i} = \sum_{i=1}^{3} \frac{0.332kPr^{1/3}\sqrt{R_x}}{x[1-(x_i/x)^{3/4}]^{1/3}} \Delta T_{w_i}$$

$$= \frac{0.332kPr^{1/3}\sqrt{R_x}}{x} \left\{ \frac{\Delta T_{w_1}}{[1-(x_1/x)^{3/4}]^{1/3}} + \frac{\Delta T_{w_2}}{[1-(x_2/x)^{3/4}]^{1/3}} + \frac{\Delta T_{w_3}}{[1-(x_3/x)^{3/4}]^{1/3}} \right\}$$

With $x_1 = 0.0$, $x_2 = 0.5$, $x_3 = 1.0$, $x = 1.5$, $\Delta T_{w_1} = 50 - 20 = 30$, $\Delta T_{w_2} = 60 - 50 = 10$, $\Delta T_{w_3} = 80 - 60 = 20$, $k = 0.028 W/m\ K$, $\nu = 18.2 \times 10^{-6} m^2/s$ (for $T_m = 50°C$) $R_x = \frac{u_e x}{\nu} = 8.2 \times 10^5$,

$$\dot{q}_w = \frac{0.332 \times 0.028 \times (0.72)^{1/3}(8.3 \times 10^5)^{1/2}}{1.5}$$

$$\left\{30.0 + \frac{10.0}{[1-(0.5/1.5)^{3/4}]^{1/3}} + \frac{20.0}{[1-(1.0/1.5)^{3/4}]^{1/3}}\right\} = 371 W/m^2$$

4.18 The integral momentum equation

$$\tau_w + \frac{dp}{dx}\delta^* = \frac{d}{dx}(\rho u_e^2 \theta) \tag{1}$$

states that the total force acting on the fluid body is equal to the rate of change of the momentum flux. The last term in (1) represents the rate of the momentum flux. Therefore, the term

$$\frac{\delta^*}{\tau_w}\left(\frac{dp}{dx}\right) \tag{2}$$

represents the ratio of the net pressure force to the wall shear force, and can be written as

$$\frac{\delta^*}{\tau_w}\frac{dp}{dx} = -\frac{\delta^*/u_e\,(du_e/dx)}{\tau_w/\rho u_e^2} \tag{3}$$

According to Thwaites' method,

$$\frac{c_f}{2} = \frac{\tau_w}{\rho u_e^2} = \frac{\nu \ell(\lambda)}{u_e \theta}, \quad \lambda = \frac{\theta^2}{\nu}\frac{du_e}{dx}, \quad H = H(\lambda) \tag{4}$$

Substituting (4) into (3) yields

$$\frac{\delta^*}{\tau_w}\frac{dp}{dx} = -\frac{H\theta^2 du_e/dx}{\nu \ell(\lambda)} = -\lambda H(\lambda)/\ell(\lambda)$$

which shows that, to the accuracy of Thwaites' method, the ratio of two net forces is uniquely related to the pressure gradient parameter $\lambda = \theta^2/\nu\ du_e/dx$.

4.19 $\delta_c^2 = \frac{x^2}{Nu_x^2} = \frac{x^2}{R_x(g_w')^2} = \frac{x\nu}{(g_w')^2 u_e}$

where g_w' is a function of Prandtl number and pressure gradient. For a fixed Pr and wedge flow, $u_e = cx^m$, g_w' is constant.

$$\therefore \quad \frac{u_e}{\nu}\frac{d\delta_c^2}{dx} = \frac{u_e}{(g_w')^2 \nu}\frac{d}{dx}\left(\frac{\nu}{c}x^{1-m}\right) = \frac{1-m}{(g_w')^2}, \quad \frac{\delta_c^2}{\nu}\frac{du_e}{dx} = \frac{x\nu}{(g_w')^2 u_e \nu} cmx^{m-1}$$

$$= \frac{m}{(g_w')^2} \quad \text{and as a result} \quad \frac{u_e}{\nu}\frac{d\delta_c^2}{dx} = \frac{1-m}{m}\frac{\delta_c^2}{\nu}\frac{du_e}{dx}$$

for wedge flows. From this relationship, the two parameters $u_e/\nu \; d\delta_c^2/dx$ and $\delta_c^2/\nu \; du_e/dx$ can be linearly related by

$$\frac{u_e}{\nu}\frac{d\delta_c^2}{dx} = A\left(\frac{\delta_c^2}{\nu}\frac{du_e}{dx}\right) + B$$

with the constants A and B depending on the Prandtl number.

4.20 a. With ξ given by (P4.26) and for a dimensionless temperature profile given by (P4.29), we can write

$$\left(\frac{\partial T}{\partial y}\right)_w = (T_e - T_w)\frac{\partial g}{\partial \xi}\frac{\partial \xi}{\partial y} = \frac{T_e - T_w}{0.893}\left(\frac{\lambda Pr}{9\nu x}\right)^{1/3}$$

$$\therefore \quad Nu_x = -\frac{x}{T_w - T_e}\left(\frac{\partial T}{\partial y}\right)_w = \frac{x}{0.893}\left(\frac{\lambda Pr}{9\nu x}\right)^{1/3}$$

b. For Blasius flow, $\lambda = \left(\frac{\partial u}{\partial y}\right)_w = 0.332\,\frac{u_e}{x}\sqrt{R_x}$

$$Nu_x = \frac{x}{0.893}\left[\frac{0.332 u_e Pr}{9\nu x^2}\sqrt{R_x}\right]^{1/3} = 0.373\, R_x^{1/2} Pr^{1/3}, \quad \text{or}$$

$$Nu_x R_x^{-1/2} = 0.373 Pr^{1/3}$$

Comparison With Exact Solution

Pr	(P4.30)	Exact Sol.	% Error
0.1	0.173	0.139	25%
1	0.373	0.332	12%
10	0.804	0.729	10%
$\to\infty$	$0.373Pr^{1/3}$	$0.339Pr^{1/3}$	10%

As can be seen from the above table, the error by the Leveque approximation decreases as the Prandtl number increases with an asymptotic value of 10% resulting from the neglect of the normal-component convective term in the energy equation whose solution is obtained with the Leveque approximation.

4.21 In (4.76a) the second term on the RHS is 0 at the stagnation point since $u_{e_1} = 0$, and $u_e = cx$ for flow near the stagnation point,

$$\therefore \quad \theta^2 = \frac{0.45\nu}{(cx)^6} \int_0^x c^5x^5dx \quad \text{or} \quad = 0.075\, x^2/R_x,$$

$$\text{or} \quad \frac{\theta}{x} = 0.27386\, R_x^{-1/2}$$

With $\lambda \equiv \frac{\theta^2}{\nu}\frac{du_e}{dx} = \frac{0.075x^2R_x^{-1}}{\nu}\frac{u_e}{x} = 0.075$ it follows from (4.78)

that $H = 2.61 - 3.75\lambda + 5.24\lambda^2 = 2.3582$,

$$\delta^* = H\theta = 0.64582xR_x^{-1/2}, \quad \frac{\delta^*}{x} = 0.64582R_x^{-1/2}$$

$$\frac{c_f}{2} = \frac{f''_w}{\sqrt{R_x}} = R_\theta^{-1}\,[0.225 + 1.61\lambda - 3.75\lambda^2 + 5.24\lambda^3]$$

$$= R_x^{-1/2}\,\frac{[0.225 + 1.61 \times 0.075 - 3.75 \times 0.075^2 + 5.24 \times 0.075^3]}{0.2783}$$

$$\therefore \quad f''_w = \sqrt{R_x}\,(\frac{c_f}{2}) = 1.19355$$

According to the Smith-Spalding method, St is given by (4.86) which, for $u_e = cx$ and with $c_2 = (c_3 - 1)/2$, can be expressed as

$$St = \frac{c_1(u_e^*)^{c_2}}{[c^{c_3}/(1 + c_3)\,(x^*)^{c_3+1}]^{1/2}}\,R_L^{-1/2} = c_1(1 + c_3)^{1/2}[R_L u_e^* x^*]^{-1/2}$$

$$= c_1(1 + c_3)^{1/2}R_x^{-1/2}$$

For $Pr = 1$, $c_1 = 0.332$ and $c_3 = 1.95$, $St = 0.5702R_x^{-1/2}$, and $g'_w = Nu_x/\sqrt{R_x} = StPrR_x/\sqrt{R_x} = 0.5702$.

The approximate and exact solutions for Pr = 1 are shown below and indicate good agreement

	Approx. Sol.	Exact Solution (Tables 4.1 & 4.3)
f''_w	1.19355	1.23259
δ^*_1	0.64582	0.64791
θ_1	0.27386	0.29234
g_w	0.5702	0.5708

4.22 For the Blasius flow, since u_e = const. and $\lambda = 0$, it follows from Thwaites' method that

$$(\frac{\theta}{L})^2 R_L = \frac{0.45}{(u_e^*)^6}\int_0^{x^*} (u_e^*)^5 dx^* = 0.45x^*$$

$$\text{or} \quad \frac{\theta}{x} = 0.6708R_x^{-1/2} \rightarrow \theta_1 = 0.6708$$

and $H = 2.61$, $\frac{\delta^*}{x} = 1.7508R_x^{-1/2} \rightarrow \delta_1^* = 1.7508$, $f_w'' = \frac{c_f}{2}\sqrt{R_x} = 0.335$

According to the Smith-Spalding method, St is given by (4.86) which, with $c_3 = 2c_2 + 1$, can be written as

$$St = \frac{c_1(u_e^*)^{c_2}}{[(u_e^*)^{2c_2+1}x^*]^{1/2}} R_L^{-1/2} = c_1 R_x^{-1/2}$$

For $Pr = 1.0$ and $c_1 = 0.332$, $St = 0.332R_x^{-1/2}$,

$$g_w' = \frac{Nu_x}{\sqrt{R_x}} = \frac{StPrR_x}{R_x^{1/2}} = 0.332$$

The approximate and exact solutions are shown below and are in excellent agreement.

	f_w''	δ_1^*	θ_1	g_w'
Thwaites or Smith-Spalding	0.335	1.7508	0.6708	0.332
Exact	0.332	1.7207	0.6641	0.332

4.23 For an axisymmetric stagnation-point flow with $u_e^* = cx^{*1/3}$ it follows from (4.76b) that

$$\left(\frac{\theta}{L}\right)^2 = \frac{0.45}{R_L(u_e^*)^6}\int_0^{x^*} c^5(x^*)^{5/3}dx^* = 0.16875(x^*)^2/R_x$$

or $\theta_1 = \frac{\theta}{x}R_x^{1/2} = 0.41079$

$$\text{and } \lambda = \frac{\theta^2}{\nu}\frac{du_e}{dx} = \frac{(0.41079xR_x^{-1/2})^2}{\nu}\frac{1}{3}\frac{u_e}{x} = 0.05625$$

$$H = 2.61 - 3.75\lambda + 5.24\lambda^2 = 2.4156, \quad \delta_1^* = \frac{\delta^*}{x}R_x^{1/2} = 0.99232$$

$$f_w'' = \frac{c_f}{2}R_x^{1/2} = \frac{R_x^{1/2}}{R_\theta}[0.225 + 1.61\lambda - 3.75\lambda^2 + 5.24\lambda^3]$$

$$= 0.30463\, R_x^{1/2}/(0.41079R_x^{1/2}) = 0.74157$$

Substituting $u_e^* = cx^{*1/3}$ into (4.86) and integrating, we get

$$St = c_1(u_e^*)^{c_2}R_L^{-1/2}\left[c^{c_3}\frac{3}{c_3+3}x^{*(c_3+3)/3}\right]^{-1/2} = c_1\left(\frac{c_3+3}{3}\right)^{1/2}R_x^{-1/2}$$

With $Pr = 1.0$, $St = 0.42646R_x^{-1/2}$, $g_w' = \frac{Nu_x}{\sqrt{R_x}} = \frac{StPrR_x}{\sqrt{R_x}} = 0.42646$

Comparison with exact solutions ($Pr = 1$) indicates good agreement, as shown by the table.

	Approx. Sol.	Exact Solution
f''_w	0.74157	0.75745
δ^*_1	0.99232	0.98536
θ_1	0.41079	0.42900
g'_w	0.42646	0.4402

4.24 In Thwaites' method, the boundary-layer parameters, θ, H, and c_f are calculated from the formulas given by (4.76b) and (4.78). Eq. (4.76b) was integrated numerically with the trapezoidal rule and $c_f/2$ and H were calculated from (4.78) for the Howarth flow, $u^*_e = 1 - ax^*$, for a = 1/8 and $R_L = 10^6$. Uniform spacing of $\Delta x^* = 0.02$ was taken between $x^* = 0.0$ and 1.0. The calculated θ/L, δ^*/L and c_f distributions are available in the diskette. From the c_f distribution given below, it is evident that the flow separates between $x^* = 0.98$ and 1.0 where either $\lambda < -0.09$ or $c_f < 0.0$.

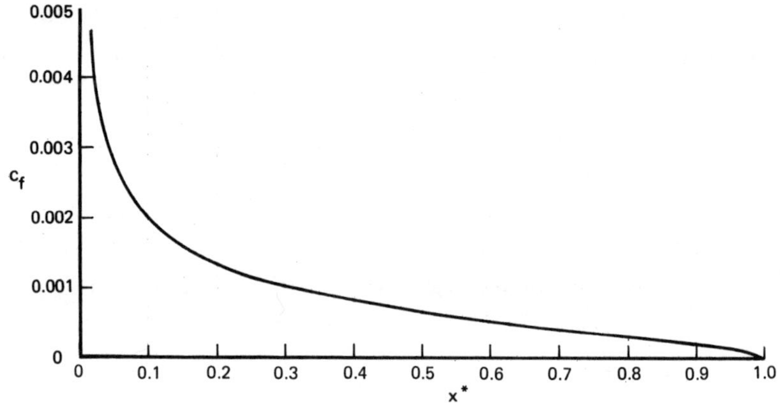

4.25 In Stratford's method, the wall shear is given by (P4.34). At the separation point, where $\tau_w = 0$, this expression reduces to

$$c_p^{1/2}\left(x \frac{dc_p}{dx}\right)^{1/2} = 0.102 \qquad (1)$$

To find the separation point, x_s, for the Howarth flow, Newton's method was used to solve (1) by first assuming $x_s = 1.0$. The calculated point of separation was $x_s = 0.972$ whereas it was between 0.98 and 1.0 with Thwaites' method. To calculate the wall shear from (P4.34) for this flow, Newton's method was again used by initially assuming τ_w/τ_β to be that of the previous point. Calculations were started at $x^* = 0.02$ for a uniform spacing of $\Delta x^* = 0.02$. The figure shows that the results of the two methods are in reasonable agreement with those obtained from the Stratford method consistently lower than those calculated by the Thwaites method discussed in Problem (P4.24). See the diskette for the computer program and its output.

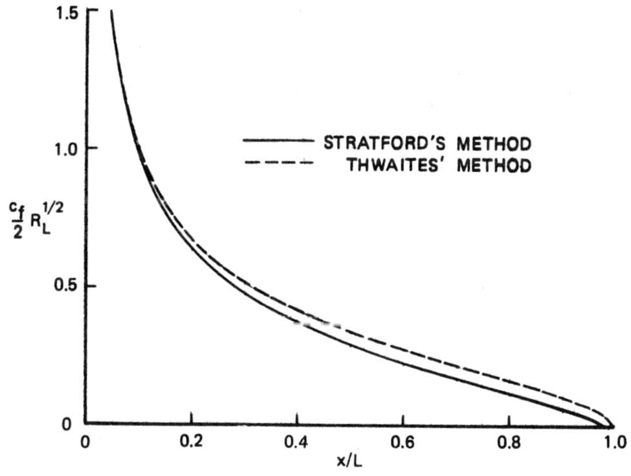

4.26 To calculate St, we use (4.86) and to calculate $c_f/2$ we use (4.76b) and (4.78). For $u_e^* = 1.0 - ax^*$, both (4.86) and (4.76b) were integrated with the trapezoidal rule for a uniform spacing of $\Delta x^* = 0.02$. The calculations were started at $x^* = 0.0$ and were continued up to $x^* = 1.0$. Separation occurred between $x^* = 0.98$ and 1.0. The distributions of the Stanton number and the ratio of St to $c_f/2$ for Pr = 1.0 and 10.0 are shown below and indicate that St increases as Pr decreases and that the Reynolds analogy $(S_t/(c_f/2) \sim 1.0)$ is valid only for Pr = 1.0 and for $x^* \ll 1.0$ where the pressure gradient is very mild and the flow is far from separation. See the diskette for the computer program and its output.

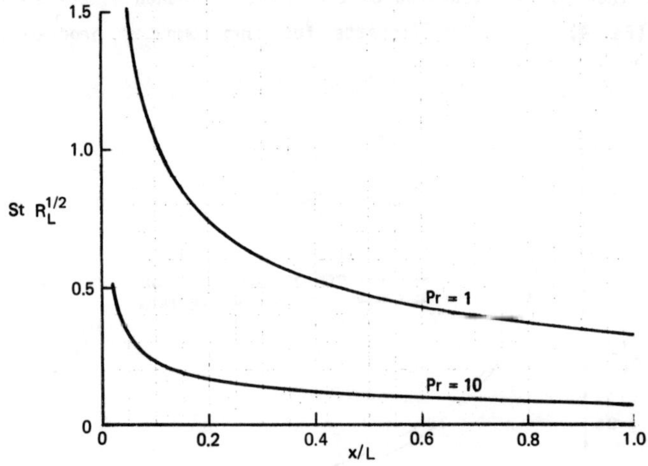

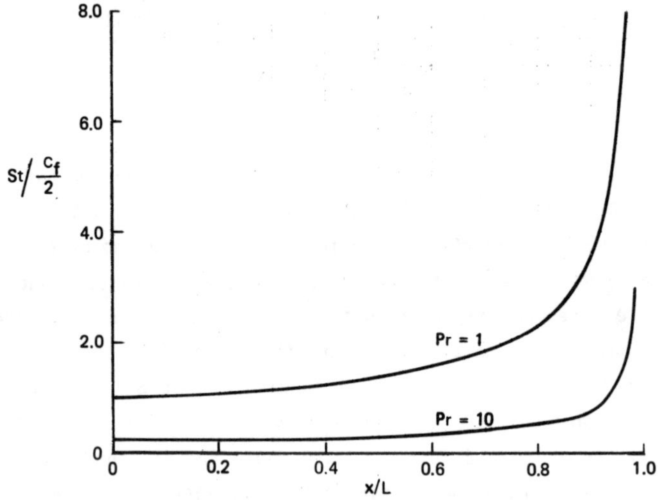

4.27 As shown in the figure, with $\beta = \tan^{-1} dy/dx$, $\theta = \pi - (\alpha + \beta)$. For a circular cylinder, $\alpha = \pi/2$ and $\theta = \pi - (\pi/2 + \beta) = \pi/2 - \beta$, and from (P4.36), $u_e = u_\infty(1 + t)\cos\beta = 2u_\infty \cos(\pi/2 - \theta) = 2u_\infty \sin\theta$.

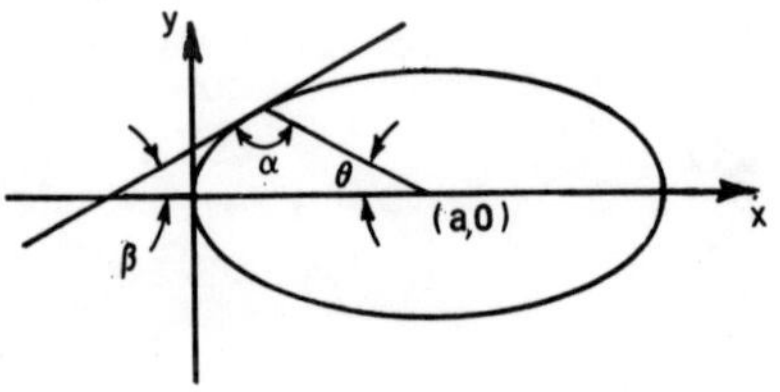

4.28 The local skin-friction coefficient, c_f, and wall heat flux rate, $\dot{q}_w$, are calculated by integrating for $u_e^* = 1.25 \cos\beta$ with A = 11.68 and B = 2.87 for Pr = 0.7. The results shown below indicate flow separation at x/a = 1.66. See the diskette for the computer program and its output.

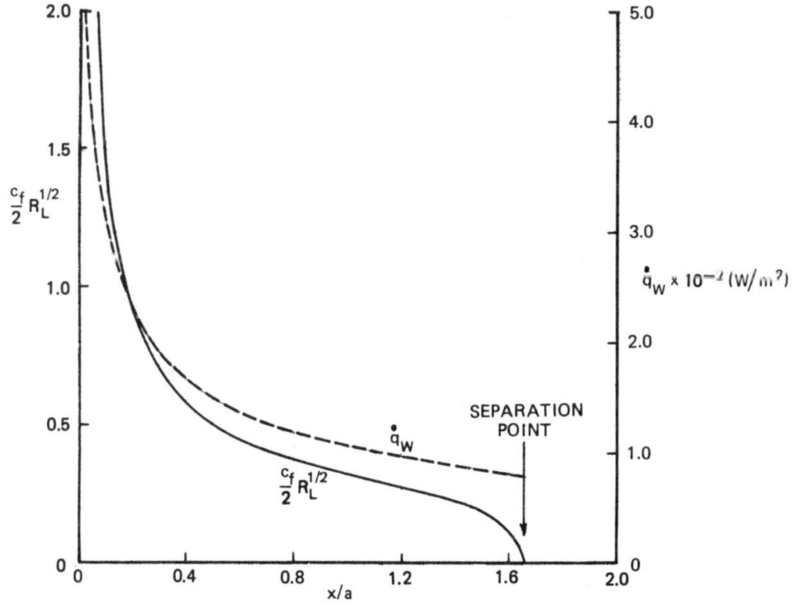

4.29 For the given body and u_e distribution, Stratford's method is applicable only for x/a > 1.0 and the calculated values of $c_f/2 \sqrt{R_L}$ are

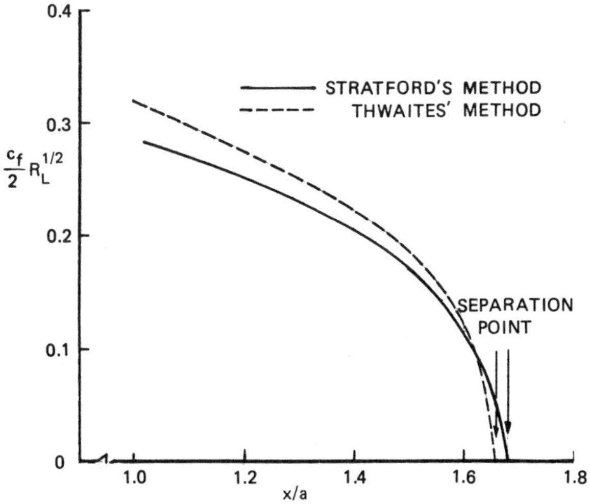

less than those computed with Thwaites' method except near the separation point which occurs at $x/a = 1.68$ by the former and 1.66 by the latter. See the diskette for the computer program and its output.

4.30 The calculated values of c_f and $\dot{q}_w$ are shown in the figure. The flow separates at about $x/a = 0.59$ where the pressure gradient is adverse and the Stanton number departs from $c_f/2$ at about $x/a = 0.15$ where the flow begins to decelerate. See the diskette for the computer program and its output.

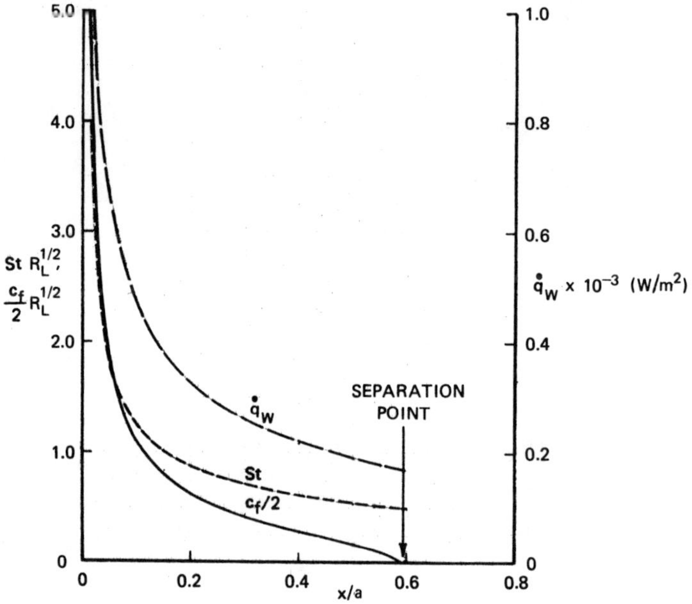

4.31 As shown in the figure, $\beta = \tan^{-1}(dr_0/dx)$, $\theta = \pi - (\alpha + \beta)$ and for a sphere, $\alpha = \pi/2$ and $\theta = \pi/2 - \beta$. To show that A, given by (P4.40), approaches 1.5 as $t \to 1.0$, we set $t^2 = 1 - \varepsilon^2$ where $\varepsilon \to 0$ as $t \to 1.0$. Then

$$(1 - t^2)^{3/2} = \varepsilon^3,$$

$$\sqrt{1 - t^2} - \frac{1}{2} t^2 \ln[(1 + \sqrt{1 - t^2})/(1 - \sqrt{1 - t^2})]$$

$$= \varepsilon - \frac{1}{2}(1 - \varepsilon^2) \ln [(1 + \varepsilon)/(1 - \varepsilon)]$$

$$= \varepsilon - \frac{1}{2}(1 - \varepsilon^2) \; [(\varepsilon - \frac{1}{2}\varepsilon^2 + \frac{1}{3}\varepsilon^3 \ldots)$$

$$- (-\varepsilon + \frac{1}{2}\varepsilon^2 - \frac{1}{3}\varepsilon^3 \ldots)] = \frac{2}{3}\varepsilon^3 + 0(\varepsilon^4)$$

$$\therefore A = \varepsilon^3/[\tfrac{2}{3}\varepsilon^3 + O(\varepsilon^4)] = 1.5 \quad \text{as} \quad \varepsilon \to 0$$

$$u_e = u_\infty A \cos\beta = 1.5u_\infty \cos(\tfrac{\pi}{2} - \theta) = 1.5u_\infty \sin\theta$$

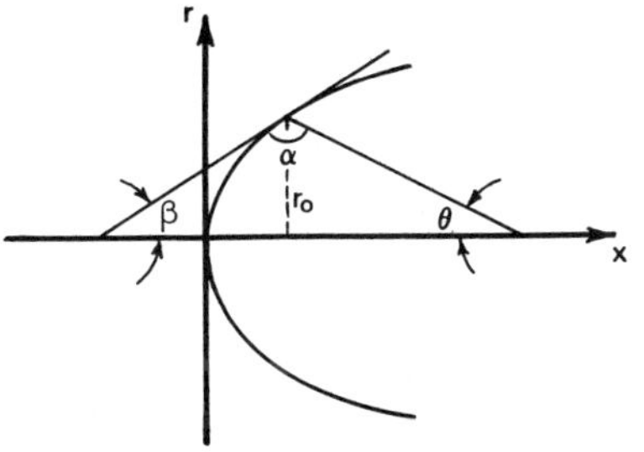

4.32 The calculated values of c_f, $\dot{q}_w$ and $St/(c_f/2)$ are plotted in the figure. The flow separates at $x/a = 1.62$ where $\lambda = -0.09$. Here x is measured along the major axis from the stagnation point. See the diskette for the computer program and its output.

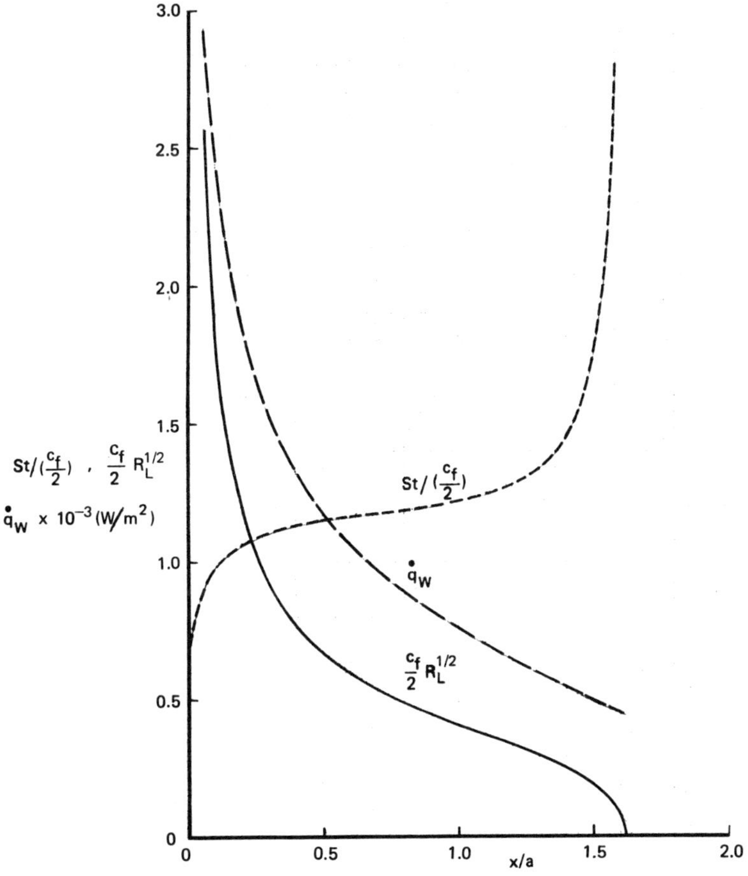

4.33 a. From Fig. 4.22 of page 111 for $u_c/u_e = 1.0$, $\delta/y_c = 1.95$ and $\eta^* = 0.5$, we have $\xi - \xi_o = 9.0$ or $x - x_o = 9.0y_c$.

b. Again, use Fig. 4.21, $f''_w = \frac{c_f}{2}\sqrt{R_x} = 0.32$

$$\therefore \quad \frac{c_f}{2} = 0.32\left[\frac{1.0 \times (0.0812 + 0.003 \times 9.0)}{1.5 \times 10^{-5}}\right]^{-1/2} = 0.00377$$

c. $\frac{\delta}{x} = \frac{\delta}{\delta^*}\frac{\delta^*}{x} = 3.08 \times 1.75R_x^{-0.5} = 5.3R_x^{-0.5}$ or $R_\delta = 5.3R_x^{0.5}$

For $\delta/y_c = 1.95$ and $y_c = 3.0$mm, $\delta = 5.85$mm,

$$R_\delta = \frac{1.0 \times 0.00585}{1.5 \times 10^{-5}} = 390, \quad R_{x_o} = \left(\frac{390}{5.3}\right)^2 = 5415.0$$

$$x_o = \frac{R_{x_o}\nu}{u_e} = 81.2\text{mm}$$

d. It will be difficult to maintain laminar flow. If laminar flow is to be achieved, x_o would become 81.2m, which would be difficult to accommodate in practice.

4.34 For $\xi - \xi_o = 16$, $u_c/u_e = 1.0$ and $\delta/y_c = 0.95$. Then, from Fig. 4.24,

$$-g'_w = 0.24 = \frac{+\dot{q}_w}{k(T_w - T_e)\sqrt{u_e/\nu x}}$$

For $\delta/y_c = 0.95$ and $y_c = 3.0$mm, $\delta = 3.0 \times 0.95 = 2.85$mm, $R_\delta = 190$

$R_{x_o} = \left(\frac{190}{5.3}\right)^2 = 1285$, $x_o = 1.93$cm

$$\therefore \quad x = 16y_c + x_o = 16 \times 0.3 + 1.93 = 4.82\text{cm}$$

$$\dot{q}_w = -g'_w k(T_w - T_e)\sqrt{u_e/\nu x} = 0.24 \times 0.026 \times 50$$

$$\times \left[\frac{1.0}{1.5 \times 10^{-5} \times 0.048}\right]^{1/2} = 368\text{W/m}^2$$

The value would decrease if δ/y_c increases to 1.95 in accord with Fig. 4.21 and Reynolds analogy.

Chapter 5

Uncoupled Laminar Duct Flows

5.1 a. The velocity profile in a circular pipe is given by (5.9) from which it is clear that the maximum velocity occurs at the center line of the pipe, i.e. $r = 0$ and is equal to

$$u_{max} = \frac{p_i - p_o}{4\mu L} r_o^2$$

The mean velocity defined by $u_m \equiv \frac{\int u dA}{A}$ then becomes

$$u_m = \frac{p_i - p_o}{4\mu L} r_o^2 \int_0^1 2 \left(\frac{r}{r_o}\right)\left[1 - \left(\frac{r}{r_o}\right)^2\right] d\left(\frac{r}{r_o}\right) = \frac{1}{2} \frac{p_i - p_o}{4\mu L} r_o^2$$

$$\therefore \quad \frac{u_m}{u_{max}} = \frac{1}{2}$$

b. Use the definition of the volume flow rate, $Q = \int u dA$, and with u given by (5.9)

$$Q = \frac{p_i - p_o}{4\mu L} r_o^2 \int_0^{r_o} 2\pi r\left[1 - \left(\frac{r}{r_o}\right)^2\right] dr = \frac{\pi r_o^4}{8\mu L} (p_i - p_o)$$

5.2 For a two-dimensional flow, $K = 0$ and (5.7) reduces to $\frac{dp}{dx} = \mu \frac{d^2u}{dy^2}$ with the boundary conditions $y = 0,\ u = 0$; $y = 2h,\ u = u_o$. Integrating this equation twice wrt y and evaluating the two integration constants from the boundary conditions, we get (P5.3).

5.3 a. With u given by (P5.4), it is obvious that the maximum velocity, u_{max}, occurs at $y = h$, $u_{max} = -(h^2/2\mu)(dp/dx)$. The mean velocity is

$$u_m = \frac{1}{A} \int u dA = \int_0^{2h} - \frac{h^2}{2\mu} \frac{dp}{dx} \frac{y}{h} \left(2 - \frac{y}{h}\right) \frac{dy}{2h} = - \frac{h^2}{3\mu} \frac{dp}{dx}$$

$$\therefore \quad \frac{u_{max}}{u_m} = \frac{3}{2}$$

b. From part (a), $-\frac{dp}{dx} = \frac{3\mu u_m}{h^2}$ or $\frac{p_i - p_o}{L} = \frac{3\mu u_m}{h^2}$

Multiply both sides of the equation by $L/\rho u_m^2$, and set $R_h = u_m h/\nu$

$$\frac{p_i - p_o}{\rho u_m^2} = \frac{3\nu L}{u_m h^2} = \frac{3L}{hR_h}$$

c. With $P \equiv \frac{-h^2}{2\mu u_0}\frac{dp}{dx} = -\frac{h^2}{2\mu u_0}\frac{p_o - p_i}{L}$ substituted into (P5.3), gives

$$\frac{u}{u_0} = \frac{y}{2h} + 4P\frac{y}{2h}(1 - \frac{y}{2h})$$

The variation of u/u_0 vs $y/2h$ for the values of P = -2, -1, 0, 1 and 2 is shown in the figure.

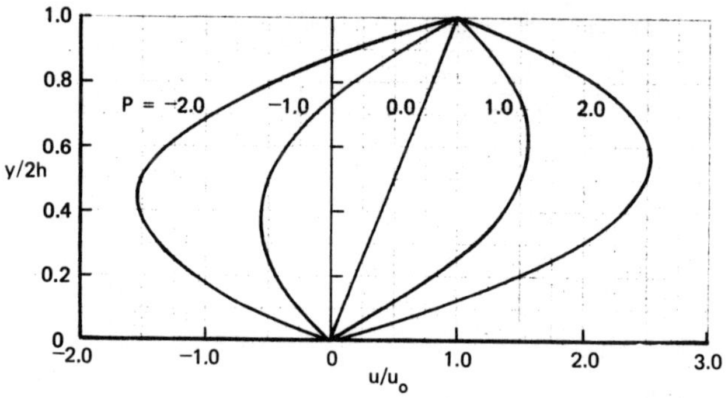

5.4 For specified wall temperature, the energy equation for the thermally fully-developed pipe flow is given by

$$uG\frac{dT_m}{dx} = \frac{\nu}{Pr}\frac{1}{r}\frac{\partial}{\partial r}(r\frac{\partial T}{\partial y}) \quad (1)$$

where $G = (T_w - T)/(T_w - T_m)$ is unknown prior to integration. In order to integrate (1), the G profile is initially assumed and updated after a new temperature profile, T, is obtained. For each temperature profile, we compute a Nusselt number and repeat the procedure until successive values of Nusselt number are such that $|Nu^{(i+1)} - Nu^{(i)}| << \varepsilon$, where ε is a tolerance parameter of, say, 10^{-4}. The computer program for performing the iterative calculations is given in the diskette. After the 7th iteration, the error in Nu is less than 10^{-4} with a converged value of 3.656.

No. of Iteration	Nu	Error, ε
1	4.36141	
2	3.72776	0.63366
3	3.66626	0.06149
4	3.65749	0.00877
5	3.65612	0.00138
6	3.65590	0.00022
7	3.65586	0.00004

5.5 The calculated Nusselt numbers for Δr = 0.1, 0.05, 0.025, and 0.01 are given below together with the exact solution.

Δr	Calculated Nu (1)	Exact Nu (2)	err (%) (1)-(2)/(2)
0.100	4.3081	4.3636	-1.27
0.050	4.3497	|	-0.32
0.025	4.3602	|	-0.08
0.010	4.3631	↓	-0.01

As expected, the error decreases as the step size Δr decreases and the solution approaches its exact value as $\Delta r \to 0$. See the diskette for the computer program and its output.

5.6 For a specified wall temperature, the normalized temperature $G(=T_w - T/T_w - T_m)$ is unknown prior to integration. The iteration procedure described in Prob. 5.4 is used to obtain the solution and the calculated Nusselt numbers for all iterations for step sizes Δr = 0.1, 0.05, 0.025 and 0.01, are tabulated below. Here ε is defined as $Nu^{(i+1)} - Nu^{(i)}$. Again, the calculated Nusselt number approaches the exact solution (= 3.658) as $\Delta r \to 0$. See the diskette for the computer program and its output.

Δr	Calculated Nu (1)	Exact Nu (2)	err (%) (1)-(2)/(2)
0.100	3.5153	3.658	-3.901
0.050	3.5994	|	-1.602
0.025	3.6315	|	-0.724
0.010	3.6475	↓	-0.287

5.7 a. Constant Heat Flux Rate

The energy equation and its boundary conditions for a 2-D fully-developed channel flow with y measured from the center are:

$$u \frac{dT_m}{dx} = \frac{\nu}{Pr} \frac{d}{dy} \left(\frac{dT}{dy}\right) \tag{1}$$

$$y = 0, \quad \frac{dT}{dy} = 0; \quad y = \pm h, \ T = T_w \tag{2}$$

The velocity profile is $u = 1.5u_m[1 - (y/h)^2]$ where u_m is the mean velocity and T_m is the mixed mean temperature

$$T_m = \int_0^h uTdh/u_m h \tag{3}$$

Integrate (1) twice wrt y and evaluate the two integration constants by satisfying boundary conditions (2); we get

$$T_w - T = \frac{3}{2}\frac{u_m Pr}{\nu} h^2 \frac{dT_m}{dx} \left[\frac{5}{12} - \frac{1}{2}\left(\frac{y}{h}\right)^2 + \frac{1}{12}\left(\frac{y}{h}\right)^4\right] \tag{4}$$

Substituting (4) into (3), $T_w - T_m = \frac{3}{2}\frac{u_m Pr}{\nu} h^2 \frac{dT_m}{dx}\frac{34}{105}$ (5)

The Nusselt number for a channel flow is defined as

$Nu = \frac{\dot{q}_w}{T_w - T_m}\frac{d_e}{k}$. Here d_e (= 4h) is the hydraulic diameter and

$$\dot{q}_w = -k\left(\frac{\partial T}{\partial y}\right)_w = k\,\frac{3}{2}\frac{u_m Pr}{\nu}\frac{dT_m}{dx} h \left(\frac{2}{3}\right) \tag{6}$$

Substituting (5) and (6) into the definition of Nu yields Nu = 140/17 = 8.235.

b. Uniform Wall Temperature

In this case, the energy equation is

$$uG\frac{dT_m}{dx} = \frac{\nu}{Pr}\frac{d}{dy}\left(\frac{dT}{dy}\right) \tag{7}$$

Initially we assume G = 1.0 which reduces (7) to (1). Eq. (7) is then integrated to obtain a temperature profile from which a new G profile is computed to integrate the energy equation again to obtain a new profile. For each temperature profile, a Nusselt number is obtained. Repeat this procedure until successive values of Nu are such that $|NU^{(i+1)} - Nu^{(i)}| < \varepsilon$, where $\varepsilon = 10^{-4}$, say. The calculated Nu and the resulting errors with this procedure are listed below.

No. Iteration	Nu	Error (ε)
1	8.23477	
2	7.58131	0.65346
3	7.54422	0.03710
4	7.54126	0.00295
5	7.54101	0.00025
6	7.54099	0.00002

After the 6th iteration, the error reduces to 2×10^{-5} and Nu becomes 7.541. These results were obtained for a uniform Δy distribution (Δy = 0.02) corresponding to 51 points from the centerline to the wall. See the diskette for the computer program and its output.

5.8 The energy equation and its boundary conditions are

$$u \frac{dT_m}{dx} = \frac{\nu}{Pr} \frac{1}{r} \frac{\partial}{\partial r} \left(r \frac{\partial T}{\partial r}\right) \tag{1}$$

$$r = 0, \quad \frac{\partial T}{\partial r} = 0; \quad r = r_0, \quad T = T_w \tag{2}$$

Integrate (1) wrt r twice and evaluate the two integration constants from b.c.'s given by (2)

$$T_w - T = \frac{Pr}{4\nu} u \frac{dT_m}{dx} (r_0^2 - r^2) \tag{3}$$

Note that u is constant across the tube. For the temperature profile given by (3), the mixed mean temperature and heat flux are,

$$T_m = T_w - \frac{Pr}{8\nu} u r_0^2 \frac{dT_m}{dx}, \quad \dot{q}_w = -k \left(\frac{\partial T}{\partial y}\right)_w = \frac{Pr}{4\nu} u \frac{dT_m}{dx} (2r_0)k$$

The Nusselt number is then

$$Nu = \frac{\dot{q}_w}{T_w - T_m} \frac{2r_0}{k} = \frac{Pr/4\nu \; u(dT_m/dx)2r_0 k}{Pr/8\nu \; ur_o^2(dT_m/dx)} \frac{2r_0}{k} = 8.0$$

5.9 The heat transfer rate per unit area is

$$\dot{q}_w = Nu(T_w - T_m) \frac{k}{d}$$

Here $Nu = 4.364$, $T_w - T_m = 5K$, $d = 0.01m$, $k = 0.03365W/m\ K$. Then $\dot{q}_w = 4.364 \times 5.0 \times 0.03365/0.01 = 73.4W/m^2$ and the heat transfer rate per unit length of the tube is

$$Q_w = \pi d \dot{q}_w = 3.14159 \times 0.01 \times 73.4 = 2.3W/m$$

5.10 For a fully-developed flow,

$$T_w - T_m = \frac{1}{Nu} \frac{Pr}{\nu} u_m r_0^2 \frac{dT_m}{dx}$$

When the wall temperature is constant,

$$\frac{d(T_m - T_w)}{T_m - T_w} = -\frac{Nu}{PrR} \frac{1}{r_0} dx$$

or $$\ln (T_{m_2} - T_w) - \ln(T_{m_1} - T_w) = -\frac{Nu}{PrR} \frac{(x_2 - x_1)}{r_0}$$

For $Nu = 3.658$, $T_{m_2} = 60°C$, $T_{m_1} = 80°C$, $T_w = 15°C$

$$Pr = \frac{\nu}{\kappa} = \frac{6 \times 10^{-6}}{6 \times 10^{-5}} = 0.1, \quad r_0 = 0.01;$$

$$R = \frac{u_m r_0}{\nu} = \frac{1.0 \times 0.01}{6 \times 10^{-6}} = 1.67 \times 10^3$$

The length required to accomplish the required cooling is then

$$\Delta x = x_2 - x_1 = \frac{1.67 \times 10^3 \times 0.1 \times 0.01}{3.658} \times [\ln(80 - 15) - \ln(60 - 15)]$$

$$= 0.168\text{m}$$

5.11 $c_p \Delta T \dot{m} = 2\pi r_o \ell \dot{q}_w, \quad \dot{q}_w = c_p \Delta T \dot{m} / 2\pi r_o \ell$

$$T_w - T_m = \frac{\dot{q}_w}{Nu} \frac{2r_o}{k} = \frac{c_p \Delta T \dot{m}}{2\pi r_o \ell} \frac{2r_o}{Nu\ k} = \frac{c_p \Delta T \dot{m}}{\pi \ell k Nu_x}$$

For $\dot{m} = 2.0\text{kg/h} = \frac{2}{3600}\ \text{kg/s}, \quad \Delta T = 50 - 10 = 40\text{K}$

$\ell = 2\text{m}, \quad k = 0.3 \times 10^{-3}\text{kW/m K}, \ c_p = 2.35\text{kJ/kg K}$

$$T_w - T_m = Nu_x^{-1} \frac{2.35 \times 40 \times 2/3600}{3.14159 \times 2.0 \times 0.3 \times 10^{-3}} = 27.7(Nu_x)^{-1}$$

With Nu_x given by Fig. 5.4, $T_w - T_m$ wrt x is given as:

$\hat{x}(= \frac{x}{r_o} \frac{1}{PrR_d})$	x(cm)	Nu_x	$T_w - T_m$(K)
0	0	∞	0
0.002	0.0027	12.00	2.31
0.004	0.0054	9.93	2.79
0.010	0.0136	7.49	3.70
0.020	0.027	6.14	4.51
0.040	0.054	5.19	5.34
0.100	0.136	4.51	6.14
∞	∞	4.36	6.35

5.12 Assume that the entrance length is much smaller than 0.6m. Then

$$u_m = -\frac{r_o^2}{8\mu}\frac{dp}{dx} = \frac{(p_i - p_o) r_o^2}{8\mu L}$$

For $p_i - p_o = 2 \text{ bars} = 2 \times 10^5\ \text{N/m}^2 = 2 \times 10^5\ \text{kg/m s}^2$,

$\mu = \rho\nu = 1300 \times 3 \times 10^{-3}\text{kg/m s} = 3.9\ \text{kg/m s}$,

$L = 0.6\text{m}, \ r_o = 0.0025\text{m}$,

$$u_m = \frac{2 \times 10^5 \times (0.0025)^2}{8 \times 3.9 \times 0.6} = 0.0668\text{m/s},$$

$$R_d = \frac{u_m d}{\nu} = \frac{0.0668 \times 0.005}{3 \times 10^{-3}} = 0.111$$

a. <u>hydrodynamic entrance length</u>

$\ell_v / r_o = 0.20R$

$\ell_v = 0.1 \times R_d \times r_o = 0.1 \times 0.111 \times 0.0025 = 2.78 \times 10^{-4}\text{m}$

$= 2.78 \times 10^{-2}\text{cm} << 0.6\text{m}$

b. <u>thermal entrance length</u>

$$Pr \equiv \frac{\rho \nu c_p}{k} = \frac{1300 \times 3 \times 10^{-3} \times 2.3}{0.3 \times 10^{-3}} = 3.0 \times 10^4$$

$$\therefore \ \ell_t = Pr\ell_v = 3.0 \times 10^4 \times 2.78 \times 10^{-4} = 8.3m$$

c. <u>the local heat transfer rate at x = 0.2, 0.4, 0.6m</u>

$$\dot{q}_w = \frac{(T_w - T_m)kNu}{d} = \frac{(30.0 - 25.0) \times 0.3 \times 10^{-3}}{0.005} Nu$$

$$= 0.3Nu \ kW/m^2$$

x	$\hat{x}$	Nu	$\dot{q}_w(kW/m^2)$
0.2	4×10^{-3}	7.9	2.37
0.4	8×10^{-3}	6.3	1.89
0.6	1.2×10^{-2}	5.3	1.59

Here $\hat{x} = \frac{x - x_o}{2r_o} \frac{1}{PrR} = 0.02 \ (x - x_o)$

5.13 $\nu = 13.8 \times 10^{-6} \ m^2/s$, $Pr = 0.712$ for air @ $T_m = 285K$,

$$R = \frac{u_m r_o}{\nu} = \frac{1 \times 0.01}{13.8 \times 10^{-6}} = 725$$

$$\ell_v = 0.20 r_o R = 0.20 \times 0.01 \times 725 = 1.45m$$

The hydrodynamic length is 1.45m whereas the heat transfer starts from 1m. Therefore the flow can be treated as hydrodynamically fully developed in calculating heat transfer parameters. The local heat transfer rate is given by

$$\dot{q}_w = \frac{(T_w - T_m)kNu}{d} = \frac{(310 - 285) \times 0.02524}{0.02} Nu = 31.5 \ Nu \ (W/m^2)$$

To determine Nu, use Fig. 5.4. The heat flux at x = 1.5m and 3.0m is:

x	$\hat{x} \ (= \frac{x - x_o}{2r_o} \frac{1}{PrR})$	Nu	$\dot{q}_w(W/m^2)$
1.5	0.048	4.2	132.3
3.0	0.291	3.658	115.2

5.14 $$T_w - T_m = \frac{\dot{q}_w d}{kNu} = \frac{25 \times 0.02}{0.02524} Nu^{-1} = \frac{19.8}{Nu}$$

x	$\hat{x}$	Nu	$T_w - T_m$	T_w
1.5	0.048	4.8	4.13	289.13
3.0	0.291	4.364	4.54	289.54

Here assume that T_m = 285K and obtain Nu from Fig. 5.4.

5.15 With the Leveque solution of the temperature distribution given by Eq. (P4.29) on p. 119, it follows that

$$\dot{q}_w = -k \frac{\partial T}{\partial y}\Big|_{y=0} = -k \frac{\partial T}{\partial \xi}\Big|_{\xi=0} \frac{\partial \xi}{\partial y} = \frac{-k}{0.893} (T_e - T_w)(\frac{\lambda Pr}{9\nu x})^{1/3}$$

where $\lambda = (\frac{\partial u}{\partial y})_{y=0} = \frac{\tau_w}{\mu} = \frac{f(1/8)\rho u_m^2}{\mu}$

For laminar flow $f = \frac{64}{R_d}$ so $\tau_w = \frac{64}{R_d} \frac{1}{8} \rho u_m^2$; as a result

$$\lambda = \frac{64}{R_d} \frac{1}{8\mu} \rho u_m^2 = \frac{64}{\rho(u_m d/\mu)} \frac{1}{8\mu} \rho u_m^2 = 8 \frac{u_m}{d}$$

$$\therefore Nu = \frac{hd}{k} = \frac{\dot{q}_w}{(T_w - T_e)} \frac{d}{k} = \frac{-k}{0.893} (T_e - T_w)(\frac{\lambda Pr}{9\nu x})^{1/3} \frac{d}{k} \frac{1}{(T_w - T_e)}$$

$$= \frac{d}{0.893} (\frac{\lambda Pr}{9\nu x})^{1/3}$$

with $\lambda = 8u_m/d$, and $x^+ = x/d$

$$Nu = \frac{1}{0.893} (\frac{8}{9} \frac{u_m d}{\nu} \frac{Pr}{x^+})^{1/3} = \frac{1}{0.893} (\frac{8}{9} \frac{Pe}{x^+})^{1/3} = 1.076 (\frac{Pe}{x^+})^{1/3}$$

Note that with x^+ defined as x/r_0 instead of x/d, and Peclet number, Pe as Pr R_d,

$$Nu = 1.355 (\frac{Pe}{x^+})^{1/3}$$

which is identical to the expression given in Problem 5.17(b).

To obtain an expression for the average Nusselt number, we integrate Nu(x) wrt x:

$$\overline{Nu} = \frac{1}{\ell_t} \int_{x=0}^{x=\ell_t} Nu(x) dx$$

and represent the thermal entrance length by

$$\frac{\ell_t}{r_0} = 0.1 \text{ Pr } R_d \quad \text{or} \quad \frac{\ell_t}{d} = \frac{\ell_t}{2r_0} = (0.05) \text{ Pe}$$

Then $\overline{Nu} = \frac{d}{\ell_t} \int_0^{\ell_t/d} Nu(x^+) \, d(\frac{x}{d}) = \frac{1}{(\ell_t/d)} \frac{1}{0.893} (\frac{8}{9})^{1/3} (Pe)^{1/3} \cdot \frac{x^{+\,2/3}}{(2/3)}\Big|_0^{\ell_t/d}$

$$= \frac{1}{0.893} (\frac{8}{9})^{1/3} (Pe)^{1/3} (\frac{3}{2}) \frac{1}{(\ell_t/d)^{1/3}} = \frac{1}{0.893} (\frac{8}{9})^{1/3} (\frac{3}{2}) Pe^{1/3}$$

$$\cdot \frac{1}{[(0.05)Pe]^{1/3}} = \frac{1}{0.893} (\frac{8}{9})^{1/3} (\frac{3}{2}) (\frac{1}{0.05})^{1/3} = 4.3766$$

Note that this value of Nusselt number is very close to the exact value of 4.364, given by Eq. (5.25). Although the assumption of linear velocity distribution is not a good one, the end result is in good agreement with the exact value. The comparison between the approximate expression $Nu(x^+) = 1.355\,(Pe/x^+)^{1/3}$ and the numerical results obtained from the computer program of Section 13.3, discussed in the solution of Problem 5.17, indicate that the predictions of both procedures are very good for $x^+ < 0.001$ Pe.

5.16 From (4.25) $T = T_e + T_e(1 - g)\phi(x)$. Then $T_m = T_e + T_e(1 - g_m)\phi(x)$

$$\therefore\quad g_m = 1 + \frac{T_m - T_e}{T_e\phi(x)} = 1 + \frac{1}{T_e\phi(x)}\left[\frac{\int_A \rho u[T_e + T_e(1 - g)\phi(x)]dA}{\rho_m u_m A} - T_e\right]$$

$$= 1 + 2\int_0^1 \hat{u}(1 - g)\hat{r}d\hat{r} \quad \text{where } \hat{u} = \frac{u}{u_m}, \quad \hat{r} = \frac{r}{r_o}, \quad \text{and with}$$

$$\hat{r} = 1 - \hat{y}, \quad d\hat{r} = -d\hat{y}$$

$$g_m = 1 + 2\int \hat{u}\,(1 - g)(1 - \hat{y})(-d\hat{y}) = 1 + 2\int_0^1 \hat{u}(g - 1)(1 - \hat{y})d\hat{y}$$

a. When the thermal layers are not merged, with $\hat{y}_e$ corresponding to the thermal-layer thickness, we have

$$g_e = \frac{T_w - T_e}{T_w - T_e} = 1.0 \quad \text{for} \quad \hat{y} > \hat{y}_e$$

$$g_m = 1 + 2\int_0^1 \hat{u}(g - 1)(1 - \hat{y})d\hat{y}$$

$$= 1 + 2\int_0^{\hat{y}_e} \hat{u}(g - 1)(1 - \hat{y})d\hat{y} + 2\int_{\hat{y}_e}^1 \hat{u}(g - 1)(1 - \hat{y})d\hat{y}$$

$$= 1 + 2\int_0^{\hat{y}_e} \hat{u}(g - 1)(1 - \hat{y})d\hat{y}$$

b. When the thermal layers are merged,

$$g_m = 1 + 2\int_0^1 \hat{u}(g - 1)(1 - \hat{y})d\hat{y}$$

$$= 1 + 2\int_0^1 \hat{u}g(1 - \hat{y})d\hat{y} - \int_0^1 \hat{u}(1 - \hat{y})d\hat{y}$$

$$= 1 + 2\int_0^1 \hat{u}g(1 - \hat{y})d\hat{y} - 1 = 2\int_0^1 \hat{u}g(1 - \hat{y})d\hat{y}$$

c. $\mathrm{Nu} = \dfrac{\dot{q}_w}{T_w - T_m}\dfrac{d}{k}$

Here $\dot{q}_w = -k\left(\dfrac{\partial T}{\partial y}\right)_w = -k/r_o\,(T_w - T_e)(-g'_w) = k/r_o\,(T_w - T_e)g'_w$

$$g_m = \frac{T_w - T_m}{T_w - T_e} \qquad \text{or} \qquad T_w - T_m = (T_w - T_e)g_m$$

$$\therefore\ \mathrm{Nu} = \frac{k}{r_o}(T_w - T_e)\,\frac{d}{k}\,\frac{1}{(T_w - T_e)g_m}\,g'_w = \frac{2g'_w}{g_m}$$

5.17 No modifications are needed for the computer program given in Section 13.3. The input to the computer program is outlined in sample calculations on page 405. Note the definition of $\hat{x}$ in this problem. The computer program uses x^+ as nondimensional x distance whereas Figure 5.4 is wrt $\hat{x}$ (i.e. $x^+ = \hat{x}/2$). The agreement between the numerical calculations and those from the analytical expressions is excellent, as shown below.

(a)

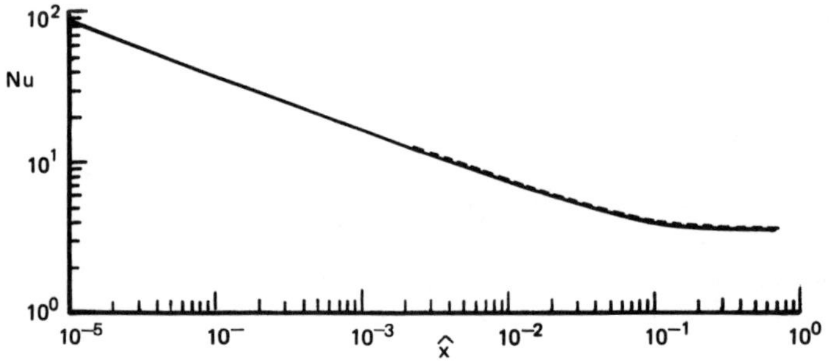

(b)

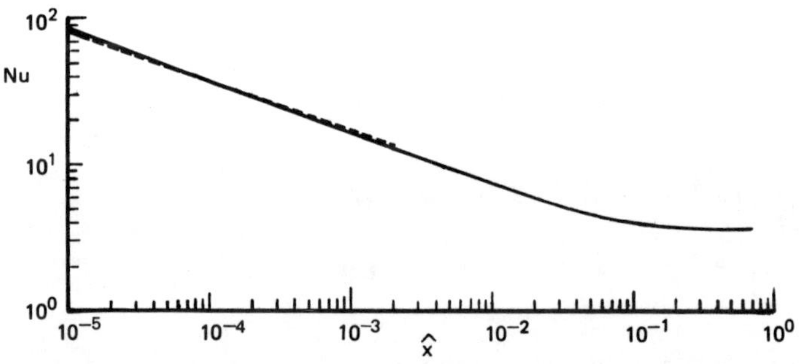

Solid line in both cases corresponds to the results of the computer program. Dashed lines indicate analytical solutions valid in the regions of their applicability.

5.18 In this case define the nondimensional temperature g by (4.25),

$$T = T_e + T_e(1 - g)\phi(\hat{x}), \quad \text{where} \quad \phi(\hat{x}) = \frac{\dot{q}_w(x)L\sqrt{\hat{x}}}{kT_e}$$

and consider the equation

$$(a_1 g')' + a_2 \frac{\eta}{2} g' - a_3 g = a_2 \hat{x} \frac{\partial g}{\partial \hat{x}} - a_3$$

before the shear layers merge subject to:

$$\eta = 0, \quad g' = 1 \quad (\text{or } p = 1), \quad \eta = \eta_e, \quad g = 1 \quad (\text{since } T = T_e)$$

where $a_1 = \hat{r}$, $a_2 = \hat{r}\hat{u}$, $a_3 = \hat{x} a_2 \hat{n}$, $\hat{n} = \frac{1}{\phi(\hat{x})} \frac{d\phi(\hat{x})}{d\hat{x}}$

For uniform wall heat fluxes, $\dot{q}_w(\hat{x})$ is constant and as a result

$$\frac{d\phi}{d\hat{x}} = \frac{\dot{q}_w}{kT_e} L \frac{1}{2} \hat{x}^{-1/2} \quad \text{or} \quad \frac{\dot{q}_w L}{2kT_e} \frac{1}{\sqrt{\hat{x}}}$$

and $$\hat{n} = \frac{1}{\phi(\hat{x})} \frac{d\phi}{d\hat{x}} = \frac{kT_e}{\dot{q}_w L \sqrt{\hat{x}}} \frac{\dot{q}_w L}{2kT_e} \frac{1}{\sqrt{\hat{x}}} = \frac{1}{2\hat{x}}$$

$$\therefore \quad a_3 = \hat{x} \cdot a_2 \cdot \hat{n} = \hat{x} \cdot a_2 \cdot \frac{1}{2\hat{x}} = \frac{a_2}{2}$$

After the shear layers merge, solve

$$(a_1 g')' - a_3 g = a_2 \frac{\partial g}{\partial \hat{x}} - a_3 \tag{5 41}$$

subject to $\hat{y} = 0$, $g' = 1$; $\hat{y} = 1$, $g' = 0$ with primes now denoting differentiation wrt $\hat{y}$ instead of η. The definitions of a_1, a_2 and $\hat{n}$ are identical to those given before the shear layers merge, except that $\hat{n} = 0$ and $a_3 = 0$ because $\phi(\hat{x})$ is constant. Note that the expression for s_3 in the computer program should be modified as

$$(s_3)_j = -\frac{1}{2}(a_3)^n_{j-1/2} - (a_2)^{n-1/2}_{j-1/2}\alpha_n \tag{13.37c}$$

In this problem it is also necessary to specify the initial temperature profile in a manner which satisfies the boundary conditions. Assume an exponentially decaying heat flux distribution in the boundary layer given by

$p(\eta,2) = \exp\left(C \frac{\eta}{\eta_e}\right)$ and compute the temperature profile from

$$g(\eta,2) = \int_0^{\eta_e} p(\eta,2)d\eta + \text{constant}$$

The integration constant is determined from $\eta = \eta_e$, $g(\eta_e) = 1.0$. In this problem the initial conditions for the heat flux and temperature profiles are assumed to be given by

$$p(\eta,2) = \exp\,(50.0 * \frac{\eta}{\eta_e}), \quad g(\eta,2) = -\frac{\eta_e}{50.0} \exp\,(\frac{50h}{\eta_e}) + 1.0$$

In the subroutine OUTPUT, the Nusselt number expression is changed to

$$Nu = \frac{2g_w}{(g_m - g_w)}$$

For further details, see the computer program in the diskette. The comparison between the numerical solutions (solid) and analytical results (dashed) are given below.

(a)

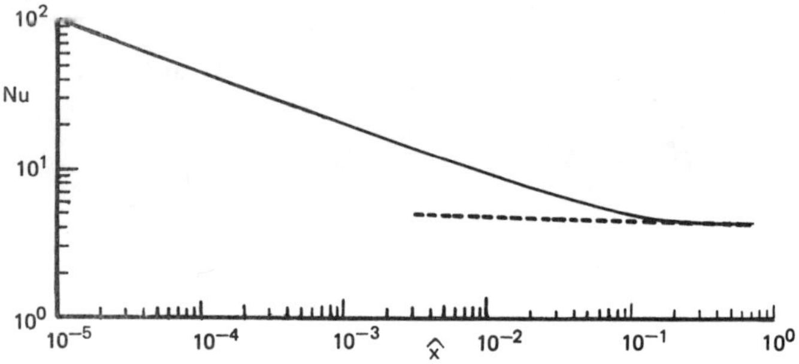

(b)

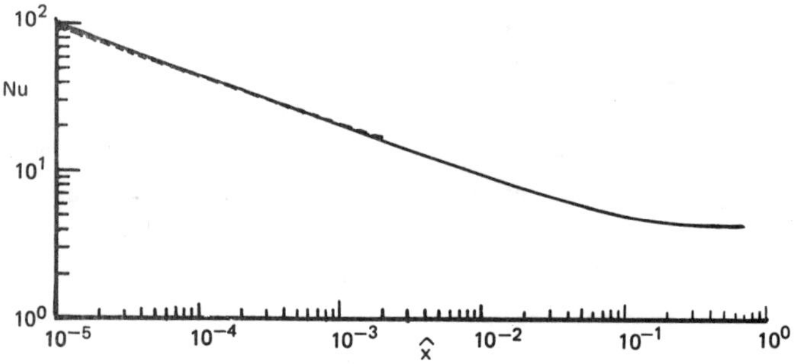

5.19 Assume a wall temperature variation given by $T_w = a + b\hat{x}$ and write it as $T_w - T_e = b\hat{x}$, since at $\hat{x} = 0$, $T_w = T_e$ and therefore $a = T_e$. With g now given by $g = (T_w - T)/(T_w - T_e)$, n is given in (5.40b) and is equal to $1/\hat{x}$. The definition of a_3 given by (5.40a) before the shear layers merge, becomes $a_3 = a_2$. After the shear layers merge, we solve (5.41) and (5.42) with the definitions of a_1 and a_2 and $\hat{n}$ which are the same as those before the merge, except that now a_3 is

$$a_3 = a_2 \frac{1}{\hat{x}} = \frac{a_2}{\hat{x}}$$

In the computer code (see the diskette) it is not necessary to modify the boundary conditions since the code is still for the specified wall temperature case. However, since there is no discontinuity at the boundary, it is not necessary to use the smoothing function. In this problem the expression for $(s_3)_j$ given by (13.37c) needs correction,

$$(s_3)_j = -\frac{1}{2}(a_3)^n_{j-1/2} - (a_2)^{n-1/2}_{j-1/2}\alpha_n$$

as well as the expression for the Nusselt number which should be written in a more general form as $Nu = 2g'_w/(g_m - g_w)$ rather than as $2g'_w/g_m$. See the computer program in the diskette for further details. The numerical (solid) and analytical results (dashed) are given below.

(a)

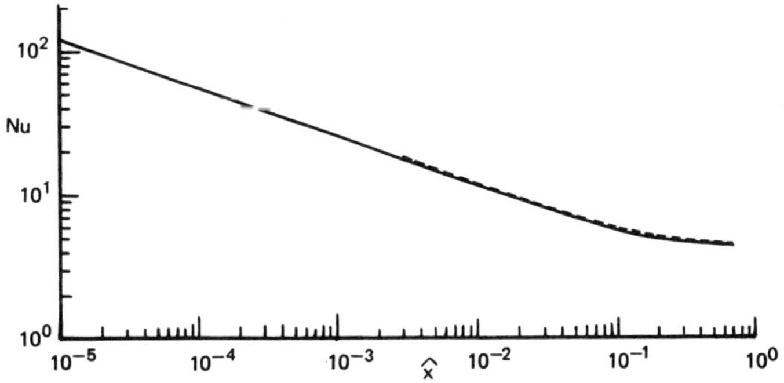

(b)

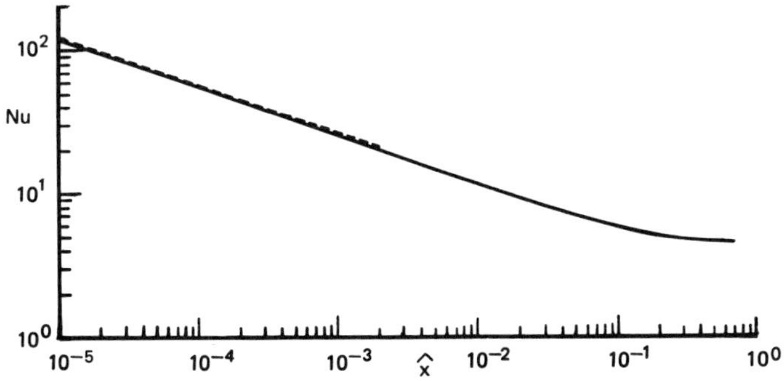

5.20 a. Let $g = \frac{T_w - T}{T_w - T_e}$, $x^* = \frac{x}{h}$, $y^* = \frac{y}{h}$, $G_z = R_h Pr = \frac{uh}{\nu} Pr$ (1)

Then the solution of (5.3)

$$G_z \frac{\partial g}{\partial x^*} = \frac{\partial^2 g}{\partial y^{*2}} \quad (2)$$

is $g(x^*,y^*) = X(x^*)Y(y^*)$ (3)

where X and Y satisfy

$$Y'' + A^2 Y = 0, \quad X' + \frac{A^2}{G_z} X = 0 \tag{4}$$

The solutions of (4) are

$$Y = a \cos Ay^* + b \sin Ay^*, \quad X = c \exp(-A^2/G_z x^*) \tag{5}$$

The constants are determined from $y^* = 0$, $Y' = 0$; $y^* = \pm 1$, $Y = 0$ to be $b = 0$, $A = (n + \frac{1}{2})\pi$ and, with $\lambda_n = (2n + 1)\pi$,

$$Y = a \cos \frac{\lambda_n}{2} y^*, \quad X = c \exp\left(- \frac{\lambda_n^2}{4} \frac{x^*}{G_z}\right)$$

Since (2) is linear, any combination of X_1 and X_2 is also a solution. Thus

$$g = \sum_{n=0}^{\infty} a_n \exp\left(- \frac{\lambda_n^2}{4} \frac{x^*}{G_z}\right) \cos \frac{\lambda_n}{2} y^*$$

where a_n are determined from $g = 1.0$ at $x^* = 0$, i.e.,

$$1 = \sum_{n=0}^{\infty} a_n \cos \frac{\lambda_n}{2} y^*$$

or $$a_n = \frac{4}{\lambda_n} \sin \frac{\lambda_n}{2} = \frac{4}{\lambda_n} \sin(n + \tfrac{1}{2})\pi = (-1)^n \frac{4}{\lambda_n}$$

$$\therefore \quad g(x^*,y^*) = \sum_{n=0}^{\infty} \frac{4}{\lambda_n} \sin \frac{\lambda_n}{2} \exp\left(- \frac{\lambda_n^2}{4} \frac{x^*}{G_z}\right) \cos \frac{\lambda_n}{2} y^*$$

$$= \sum_{n=0}^{\infty} (-1) \frac{4}{\lambda_n} \exp\left(- \frac{\lambda_n^2}{4} \frac{x^*}{G_z}\right) \cos \frac{\lambda_n}{2} y^*$$

b. $$\dot{q}_w = -k \left(\frac{\partial T}{\partial y}\right)_w = -k(T_e - T_w) \frac{1}{h} g'(-1)$$

$$= -k(T_e - T_w)/h \sum_{n=0}^{\infty} 2 \exp\left(- \frac{\lambda_n^2}{4} \frac{x^*}{G_z}\right)$$

$$T_w - T_m = \int_0^1 (T_w - T)dy^* = (T_w - T_e) \int_0^1 g dy^*$$

$$\therefore \quad Nu_x = \frac{\dot{q}_w}{T_w - T_m} \frac{x}{k} = \frac{2x^* \sum_{n=0}^{\infty} \exp\left(- \frac{\lambda_n^2}{4} \frac{x^*}{G_z}\right)}{\int_0^1 g dy^*}$$

c. See the diskette for the computer program.

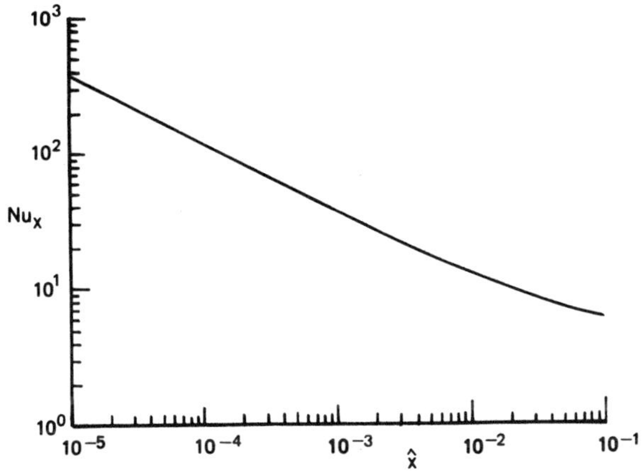

5.21 a. In transformed variables, we define

$$d\eta = \left(\frac{u_0}{\nu x}\right)^{1/2} \frac{r}{L}\, dy, \quad \eta_{sp} = \int_0^{r_0} \left(\frac{u_0}{\nu x}\right)^{1/2} \frac{r_0 - y}{L}\, dy = \left(\frac{u_0}{\nu x}\right)^{1/2} \frac{r_0^2}{2L} \tag{1}$$

$$u_0 \pi r_0^2 = \int_0^{r_0} 2\pi r u dr = \int_0^{\eta_{sp}} 2\pi u \left(\frac{\nu x}{u_0}\right)^{1/2} L d\eta$$

$$\therefore \quad \int_0^{\eta_{sp}} \frac{u}{u_0}\, d\eta = \frac{1}{2} \frac{r_0^2}{L} \left(\frac{u_0}{\nu x}\right)^{1/2} = f(x, \eta_{sp}) \tag{2}$$

From (1) and (2), we then have $f(x,\eta_{sp}) = \eta_{sp}$

b. In primitive variables, with $R_L = u_0 r_0/\nu$, we define

$$dY = \left(\frac{u_0}{\nu L}\right)^{1/2} \left(\frac{r}{L}\right) dy, \quad Y_c = \int_0^{r_0} \left(\frac{u_0}{\nu L}\right)^{1/2} \frac{r_0 - y}{L}\, dy = \frac{1}{2}\sqrt{R_L}$$

$$u_0 \pi r_0^2 = \int_0^{r_0} 2\pi r u dr = \int_0^{Y_c} 2\pi u \left(\frac{\nu L}{u_0}\right)^{1/2} L dY$$

$$\therefore \quad F(x, Y_c) = \int_0^{Y_c} \frac{u}{u_0}\, dY = \frac{1}{2} \left(\frac{u_0}{\nu L}\right)^{1/2} \frac{r_0^2}{L} = \frac{1}{2}\sqrt{R_L}$$

Note that $r_0 = L$.

5.22 a. For a circular duct, with $R_L = u_0 r_0/\nu$,

$$dY = \left(\frac{u_0}{\nu L}\right)^{1/2} \frac{r}{L}\, dy,$$

$$Y_c = \int_0^{r_0} \left(\frac{u_0}{\nu L}\right)^{1/2} \frac{r_0 - y}{L}\, dy = \left(\frac{u_0}{\nu L}\right)^{1/2} \frac{r_0^2}{2L} = \frac{1}{2}\sqrt{R_L}$$

b. For a plane duct, with $R_L = (\frac{u_o h}{\nu})$

$$dY = (\frac{u_o}{\nu L})^{1/2} dy, \quad Y_c = \int_0^h (\frac{u_o}{\nu L})^{1/2} dy = (\frac{u_o}{\nu L})^{1/2} h = \sqrt{R_L}$$

5.23 Along the centerline, F = constant or $\partial F/\partial x = 0$, and

$$(bF")' = \{(1 - \frac{2y}{\sqrt{R_L}})F"\}'_{y=y_c} = \{1 - \frac{2y}{\sqrt{R_L}} F"' - \frac{2}{\sqrt{R_L}} F"\}_{y=y_c} = - \frac{2}{\sqrt{R_L}} F"$$

$$(eg')' = - \frac{1}{Pr} \frac{2}{\sqrt{R_l}} g'$$

$$\therefore \quad - \frac{2}{\sqrt{R_L}} F"_c = \frac{dp^*}{dx} + F'_c \frac{dF'_c}{dx}, \quad F"_c = - \frac{1}{2} \sqrt{R_L} \{\frac{dp^*}{dx} + \frac{1}{2} \frac{dF'^2_c}{dx}\}$$

$$g'_c = - \frac{1}{2} \sqrt{R_L} \, PrF'_c \{n(g_c - 1) + \frac{dg_c}{dx} \}$$

Chapter 6

Uncoupled Turbulent Boundary Layers

6.1 As a first step we eliminate the dimension [M] by dividing all the variables containing [M] by ρ

	L	T	θ
$T_w - T$	0	0	1
y	1	0	0
τ_w/ρ	2	-2	0
$\dot{q}_w/\rho$	3	-3	0
μ/ρ	2	-1	0
k/ρ	4	-3	-1
c_p	2	-2	-1

Next, eliminate the dimension [T] by selecting τ_w/ρ as a dividend based on our physical judgment

	L	θ
$T_w - T$	0	1
y	1	0
$\dot{q}_w/\rho\ (\tau_w/\rho)^{-3/2}$	0	0
$\mu/\rho\ (\tau_w/\rho)^{-1/2}$	1	0
$k/\rho\ (\tau_w/\rho)^{-3/2}$	1	-1
$c_p\ (\tau_w/\rho)^{-1}$	0	-1

Now we eliminate the dimension of [θ]

	L	θ
$c_p(T_w - T)(\tau_w/\rho)^{-1}$	0	0
y	1	0
$\dot{q}_w/\rho(\tau_w/\rho)^{-3/2}$	0	0
$\mu/\rho\ (\tau_w/\rho)^{-1/2}$	1	0
$k/\rho c_p\ (\tau_w/\rho)^{-1/2}$	1	0

Finally, by inspection, the length scale can be eliminated by forming the groups, $(\tau_w/\rho)^{1/2}\rho y/\mu$, $\mu c_p/k$

$$\frac{c_p(T_w - T)}{\tau_w/\rho} = f\left[\frac{\dot{q}_w}{\rho}\left(\frac{\tau_w}{\rho}\right)^{-3/2}, \left(\frac{\tau_w}{\rho}\right)^{1/2}\frac{\rho y}{\mu}, \frac{\mu c_p}{k}\right]$$

which can be rewritten as

$$\frac{T_w - T}{\dot{q}_w/[\rho c_p(\tau_w/\rho)^{1/2}]} = f\left[\frac{\dot{q}_w}{\rho}\left(\frac{\tau_w}{\rho}\right)^{-3/2}, \left(\frac{\tau_w}{\rho}\right)^{1/2}\frac{\rho y}{\mu}, \frac{\mu c_p}{k}\right]$$

Introduce the definitions,

$$u_\tau \equiv \left(\frac{\tau_w}{\rho}\right)^{1/2}, \quad y^+ \equiv \frac{y\rho u_\tau}{\mu}, \quad Pr \equiv \frac{\mu c_p}{k}$$

so that $\dfrac{T_w - T}{\dot{q}_w/\rho c_p u_\tau} = f\left(\dfrac{\dot{q}_w}{\rho u_\tau^3}, y^+, Pr\right)$

6.2 For the flow on a rough surface, the relevant variables describing the velocity profile near the wall are u, ρ, μ, y, τ_w and k, where k is a measure of the roughness height. With the matrix elimination method, we construct a matrix whose column gives mass, length, and time dimensions of the variables in each row

	M	L	T
u	0	1	-1
y	0	1	0
τ_w	1	-1	-2
μ	1	-1	-1
ρ	1	-3	0
k	0	1	0

We first eliminate the dimension [M] by dividing the variables containing the dimension [M] by ρ and eliminate the dimension [T].

	L	T
$u(\tau_w/\rho)^{-1/2}$	0	0
y	1	0
$\mu/\rho\ (\tau_w/\rho)^{-1/2}$	1	0
k	1	0

Finally, by inspection, the length scale is eliminated by forming the groups $y(\tau_w/\rho)^{1/2}/(\mu/\rho)$, $k(\tau_w/\rho)^{1/2}/(\mu/\rho)$.

$$f\left[\frac{u}{(\tau_w/\rho)^{1/2}}, \frac{y(\tau_w/\rho)^{1/2}}{\mu/\rho}, \frac{k(\tau_w/\rho)^{1/2}}{\mu/\rho}\right] = 0$$

or $u^+ = \dfrac{u}{u_\tau} = g(y^+, k^+)$ where $u_\tau \equiv \left(\dfrac{\tau_w}{\rho}\right)^{1/2}$, $y^+ = \dfrac{yu_\tau}{(\mu/\rho)}$, $k^+ = \dfrac{ku_\tau}{(\mu/\rho)}$

6.3 In the viscous sublayer, $u^+ = y^+$ or $u = u_\tau^2/\nu\, y$. From the continuity equation,

$$v = -\int \frac{\partial u}{\partial x}\, dy = \frac{u_\tau}{\nu}\frac{du_\tau}{dx}\, y^2$$

so that $u \cdot v \sim y^3$ and $-\overline{u'v'} \sim \sqrt{\overline{u'^2}}\sqrt{\overline{v'^2}} = O(uv) = O(y^3)$ near the wall. Thus $-\overline{u'v'}$ should vary, at least, as the third power of y. From the Van Driest formula for the mixing length,

$$-\overline{u'v'} = \{\kappa y[1 - \exp(-y^+/A^+)\}^2\, (\partial u/\partial y)^2$$

In the viscous sublayer, $\partial u/\partial y \sim u_\tau^2/\nu$, and

$$1 - \exp(-y^+/A^+) \sim 1.0 - [1.0 - y^+/A^+ \ldots] \sim y$$

so $-\overline{u'v'} \sim y^4$

6.4 $$\frac{(\gamma - 1)(u_\tau/a_w)^2}{\dot{q}_w/(\rho c_p u_\tau T_w)} = \frac{\rho u_\tau^3\, (\gamma - 1)/a_w^2}{\dot{q}_w/(c_p T_w)} = \frac{\rho u_\tau^3}{\dot{q}_w}\,\frac{(\gamma - 1)c_p T_w}{\gamma R T_w}$$

$$(\because a_w = \sqrt{\gamma R T_w}) = \frac{\rho u_\tau^3}{\dot{q}_w} \quad \text{since } c_p = \frac{\gamma R}{\gamma - 1}$$

6.5 The momentum equation near the wall of a flat plate with a transpiration velocity, v_w, is

$$v_w \frac{\partial u}{\partial y} = \frac{\partial \tau}{\partial y} \quad \text{so} \quad \tau = \tau_w + v_w u$$

Then from (6.13),

$$\frac{\partial u}{\partial y} = \frac{(\tau_w + v_w u)^{1/2}}{\kappa y}, \qquad \frac{du}{(\tau_w + v_w u)^{1/2}} = \frac{dy}{\kappa y}$$

After integration, $\frac{2}{v_w}\{(u_\tau^2 + v_w u)^{1/2} - u_\tau\} = \frac{1}{\kappa}\ln y + c' = \frac{1}{\kappa}\ln y^+ + c$ where c is a function of v_w in general.

6.6 When v_w is small, $(u_\tau^2 + v_w u)^{1/2} = u_\tau\left(1 + \frac{v_w u}{u_\tau^2}\right)^{1/2} \simeq u_\tau\left[1 + \frac{1}{2}\frac{v_w u}{u_\tau^2}\right]$

$$\frac{2}{v_w}[(u_\tau^2 + v_w u)^{1/2} - u_\tau] \simeq \frac{2}{v_w}\left[u_\tau + \frac{1}{2}\frac{v_w u}{u_\tau} - u_\tau\right] = \frac{u}{u_\tau}$$

so, as $v_w \to 0$, $\frac{u}{u_\tau} = \frac{1}{\kappa}\ln y^+ + c$

6.7 For the velocity profile given by

$$\frac{u}{u_\tau} = \frac{1}{\kappa}\ln\frac{y u_\tau}{\nu} + c + \frac{\Pi}{\kappa}\left[1 - \cos\pi\frac{y}{\delta}\right], \tag{1}$$

we can write $\frac{u_e - u}{u_\tau} = -\frac{1}{\kappa}\ln\frac{y}{\delta} + \frac{\Pi}{\kappa}[1 + \cos\pi\frac{y}{\delta}]$. Then

$$\delta^* = \int_0^\delta (1 - \frac{u}{u_e})dy = \frac{\delta u_\tau}{u_e}\int_0^\delta \frac{u_e - u}{u_\tau} d(\frac{y}{\delta})$$

$$= \frac{\delta u_\tau}{u_e}\int_{e\to 0}^{1} [-\frac{1}{\kappa}\ln\eta + \frac{\Pi}{\kappa}(1 + \cos\pi\eta)]d\eta \quad = \frac{\delta u_\tau}{\kappa u_e}(1 + \Pi) \tag{2}$$

$$\theta = \int_0^\delta \frac{u}{u_e}(1 - \frac{u}{u_e})dy = \frac{\delta u_\tau^2}{u_e^2}\int_0^\delta [\frac{u_e}{u_\tau}(\frac{u_e - u}{u_\tau}) - (\frac{u_e - u}{u_\tau})^2]d(\frac{y}{\delta})$$

After some rearranging, we have

$$\kappa^2 \frac{(\delta^* - \theta)u_e^2}{\delta u_\tau^2} = 2 + 2[1 + \frac{1}{\pi}Si(\pi)]\Pi + \frac{3}{2}\Pi^2 \tag{3}$$

where $Si(\pi) = \int_0^\pi \frac{\sin u}{u} du$. Divide (2) by (3),

$$\frac{\delta^* u_\tau}{\kappa(\delta^* - \theta)u_e} = \frac{1 + \Pi}{2 + 2[1 + 1/\pi\ Si(\pi)]\Pi + 3/2\ \Pi^2} \equiv F(\Pi) \tag{4}$$

and from the definition of $H = \delta^*/\theta$, it follows from (4) that

$$\frac{Hu_\tau}{\kappa(H - 1)u_e} \equiv \frac{1}{\kappa G} = F(\Pi) \quad \text{where} \quad G = \frac{(H - 1)u_e}{Hu_\tau}$$

6.8 For a flat-plate boundary layer with high Reynolds number, say, $R_\theta > 5000$, $\Pi = 0.55$ and from (P6.3),

$$\frac{\delta u_\tau}{\nu} = \frac{\kappa}{1 + \Pi}\frac{\delta^* u_e}{\nu} = \frac{0.41}{1 + 0.55} \times 20{,}000 = 5290$$

From (P6.2), $(\frac{2}{c_f})^{1/2} = \frac{1}{\kappa}\ln\frac{\delta u_\tau}{\nu} + c + \frac{2\Pi}{\kappa}$

$$= \frac{1}{0.41}\ln 5290 + 5.0 + \frac{2 \times 0.55}{0.41} = 28.59$$

$\therefore\ c_f = 2.45 \times 10^{-3}$

For $\Pi = 0.55$, $F(\Pi) = \frac{1 + 0.55}{2 + 2(1.0 + 1.8519/\pi) \times 0.55 + 1.5 \times 0.55^2} = 0.369$

From (P6.5), $\frac{H}{H - 1} = \kappa\sqrt{2/c_f}\ F(\Pi) = 0.41 \times 28.59 \times 0.369 = 4.325$

$\therefore\ H = 1.301$ and $R_\theta = \frac{R_{\delta^*}}{H} = \frac{20{,}000}{1.301} = 15{,}370$

6.9 a. $\nu = 1.5 \times 10^{-5}\ m^2/s$ for air at $T = 25°C$. For Blasius flow, $\theta/x = 0.664/\sqrt{R_x}$; At transition point, where $R_{x_{tr}} = 3 \times 10^6$,

$$x_{tr} = \frac{R_{x_{tr}}\,\nu}{u_e} = \frac{3 \times 10^6 \times 1.5 \times 10^{-5}}{30.0} = 1.5\text{m}$$

$$\theta_{tr} = x_{tr}\frac{0.664}{\sqrt{R_{x_{tr}}}} = 1.5 \times 0.664 \times (3 \times 10^6)^{-1/2} = 0.575 \times 10^{-3}\text{m}$$

Assuming that the momentum thickness is continuous at transition, and with x_0 representing the effective origin of turbulent flow (see the figure), we have

$$\frac{\theta_{tr}}{(x_{tr} - x_0)} = 0.036\left[\frac{u_e(x_{tr} - x_0)}{\nu}\right]^{-0.20}$$

$$(x_{tr} - x_0)^{0.80} = \frac{0.575 \times 10^{-3}}{0.036}\left[\frac{30.0}{1.5 \times 10^{-5}}\right]^{0.2} = 0.2908$$

$\therefore \quad x_0 = 1.286\text{m}$

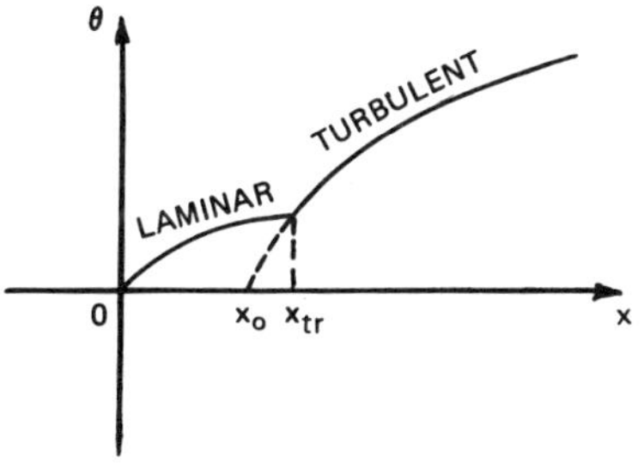

b. $$c_f = 0.059\,R_x^{-0.20} = 0.059\left[\frac{30 \times (5.0-1.286)}{1.5 \times 10^{-5}}\right]^{-0.20} = 2.493 \times 10^{-3}$$

$$\overline{c_f} = \left[\int_0^{x_{tr}} (c_f)_\ell dx + \int_{x_{tr}}^{\ell} (c_f)_t dx\right] / \ell = (L + T)/\ell$$

$$L = \int_0^{x_{tr}} (c_f)_\ell dx = \int_0^{x_{tr}} \frac{0.664}{\sqrt{R_x}} dx = 1.328\, x_{tr}/\sqrt{R_{x_{tr}}}$$

$$T = \int_{x_{tr}}^{\ell} (c_f)_t dx = 0.059 \int_{x_{tr}}^{\ell}\left[\frac{u_e(x - x_0)}{\nu}\right]^{-0.20} dx$$

$$= 0.074\,(x - x_0)\left[\frac{u_e(x - x_0)}{\nu}\right]^{-0.20}\Bigg|_{x=x_{tr}}^{x=\ell}$$

For $x_{tr} = 1.5$m, $x_0 = 1.286$m, $\ell = 5$m, $u_e = 30$m and $\nu = 1.5 \times 10^{-5}\text{m}^2/\text{s}$,

$$L = 1.3289 \times 1.5 \times \left[\frac{30.0 \times 1.5}{1.5 \times 10^{-5}}\right]^{-1/2} = 1.15 \times 10^{-3}$$

$$T = 0.074 \times \left(\frac{30.0}{1.5\times10^{-5}}\right)^{-0.2} [(5.0 - 1.286)^{0.8} - (1.5 - 1.286)^{0.8}]$$

$$= 1.057 \times 10^{-2}$$

$$\therefore \bar{c}_f = (L + T)/\ell = \frac{(1.057 + 0.115) \times 10^{-2}}{5.0} = 2.137 \times 10^{-3} \qquad (1)$$

$\bar{c}_f$ can also be calculated from (6.60) or (6.61), where the constant depends on $R_{x_{tr}}$. For $R_{x_{tr}} = 3 \times 10^6$, $A = 8700$ so that $\bar{c}_f = 2.134 \times 10^{-3}$ from (6.60) and $\bar{c}_f = 2.076 \times 10^{-3}$ from (6.61). Both results agree well with that given by (1).

c. At $x = 4.0$m

$$c_f = 0.059 \left[\frac{30.0 \times (4.0 - 1.286)}{1.5 \times 10^{-5}}\right]^{-0.20} = 2.654 \times 10^{-3}$$

$$u_\tau = u_e \sqrt{c_f/2} = 30.0 \left[\frac{2.654 \times 10^{-3}}{2.0}\right]^{1/2} = 1.093 \text{ m/s}$$

$$y = \frac{y^+\nu}{u_\tau} = \frac{1.5 \times 10^{-5}}{1.093} y^+$$

For $y^+ = 5, 50, 100, 500$ and 1000, y are 6.86×10^{-3}, 6.86×10^{-2}, 1.37×10^{-1}, 0.686 and 1.37 cm, respectively.

6.10 a. From Appendix B, $\rho = 1.087$ kg/m^3, $c_p = 1007$ J/kg K, $\nu = 1.8 \times 10^{-5}$ m^2/s, $Pr = 0.702$ at $T_f = 1/2\,(T_w + T_e) = 50°$C.

At transition, $R_x = 3 \times 10^6$ so $x_{tr} = R_x\nu/u_e = 1.8$m. To determine the effective origin, x_0, assume θ to be continuous at transition.

$$0.664\, x_{tr} R_{x_{tr}}^{-1/2} = 0.036(x_{tr} - x_0)\left[\frac{u_e(x_{tr} - x_0)}{\nu}\right]^{-0.20}$$

For $x_{tr} = 1.8$m and $R_{x_{tr}} = 3 \times 10^6$, $x_0 = 1.544$m.

The rate of cooling of the plate per unit width is given by

$$Q_w = \int_0^\ell \dot{q}_w dx = \int_0^{x_{tr}} (\dot{q}_w)_\ell dx + \int_{x_{tr}}^{\ell} (\dot{q}_w)_t dx$$

$$= \rho c_p \nu (T_w - T_e) \left[\int_0^{R_{x_{tr}}} (St)_\ell dR_x + \int_{R_{x_{tr}}}^{R_\ell} (St)_t dR_x\right]$$

For $R_x < R_{x_{tr}}$, $St = 0.332\, Pr^{-2/3} R_x^{-1/2}$

$$\text{and} \int_0^{R_{x_{tr}}} St\,dR_x = 0.664 Pr^{-2/3} R_{x_{tr}}^{1/2}$$

For $R_x > R_{x_{tr}}$ use the Reynolds analogy

$$St = \frac{c_f}{2} = 0.0296 R_{\tilde{x}}^{-0.20} \text{ with } R_{\tilde{x}} = u_e(x - x_0)/\nu \text{ so that}$$

$$\int_{R_{x_{tr}}}^{R_\ell} St\,dR_x = 0.037\,[(R_\ell - R_{x_0})^{0.80} - (R_{x_{tr}} - R_{x_0})^{0.80}]$$

$$\therefore \quad Q_w = \rho c_p \nu (T_w - T_e)\{0.664 Pr^{-2/3} R_{x_{tr}}^{1/2} + 0.037[(R_\ell - R_{x_0})^{0.80} - (R_{x_{tr}} - R_{x_0})^{0.80}]\}$$

$$= 1.087 \times 1007 \times 1.8 \times 10^{-5} \times (75 - 25)\{0.664 \times (0.702)^{-2/3}(3 \times 10^6)^{1/2} + 0.037[(8.33 \times 10^6 - 2.57 \times 10^6)^{0.80} - (3.0 \times 10^6 - 2.57 \times 10^6)^{0.80}]\}$$

$$= 9597 \text{ W/m}$$

b. If the flow is turbulent from the leading edge, then

$$Q_w = \rho c_p \nu (T_w - T_e)[0.037 R_\ell^{0.80}] = 12.53 \times 10^3 \text{ W/m}$$

The error is $\frac{12.53 - 9.597}{9.597} \times 100\% = 30.56\%$

i.e., the cooling rate will be overestimated by 31%.

c. For $u_e = 15$m, $x_{tr} = 3.6$m and $x_0 = 3.088$m

$$Q_w = 3.073 \times 10^3 \text{ W/m} \quad \text{with} \quad R_{x_{tr}} = 3 \times 10^6$$

and $Q_w = 7.204 \times 10^3$ W/m with $R_{x_{tr}} = 0$

and the error is $\frac{7.204 - 3.074}{3.074} \times 100\% = 134\%$, more than 100%, as compared with 31% in part (b). The results of parts (b) and (c) clearly indicate the importance of the transition location in estimating the heating (or cooling) rate and indicate that the error due to transition location decreases with increasing Reynolds number.

6.11 From Appendix B, Table B-2,

$\rho = 663$ kg/m^3, $c_p = 4518.7$ J/kg K, $\nu = 3.82 \times 10^{-7}$m^2/s, $Pr = 2.07$ for liquid ammonia at $T_f = 1/2(T_w + T_e) = (-10 - 25)/2 = -17.5$. At transition $R_{x_{tr}} = 4 \times 10^5$. Then

$4 \times 10^5 = 2x_{tr}/(3.82 \times 10^{-7})$, so that, $x_{tr} = 0.076$m, $x_o = 0.053$m

The local heat transfer coefficient $\hat{h}$ at the end of the plate is

$$\hat{h} = \frac{\dot{q}_w}{T_w - T_e} = \rho c_p u_e St = \rho c_p u_e \, [0.0296 R_{\tilde{x}}^{-0.20}]$$

$$= 663 \times 4518.7 \times 2.0 \times 0.0296 \, \left[\frac{2.0 \times (2.0 - 0.053)}{3.82 \times 10^{-7}}\right]^{-0.20}$$

$$= 7033.7 \text{ W/m}^2 \text{ K}$$

and the average heat transfer coefficient is

$$\hat{h}_m = \int_0^{\ell} \hat{h}\, dx/\ell = \rho \nu c_p \left\{ \int_0^{R_{x_{tr}}} St dR_x + \int_{R_{x_{tr}}}^{R_\ell} St dR_x \right\}/\ell$$

$$= \frac{\rho \nu c_p}{\ell} \{0.664 Pr^{-2/3} R_{x_{tr}}^{1/2} + 0.037[(R_\ell - R_{x_o})^{0.80} - (R_{x_{tr}} - R_{x_o})^{0.80}]\}$$

$$= 663 \times 4518.7 \times 3.82 \times 10^{-7} \{0.664 \times (2.07)^{-2/3} (4 \times 10^5)^{1/2}$$

$$+ 0.037[(1.047 \times 10^7 - 2.8 \times 10^5)^{0.8}$$

$$- (4 \times 10^5 - 2.8 \times 10^5)^{0.80}]\}/2.0$$

$$= 663 \times 4518.7 \times 3.82 \times 10^{-7} \{258.6 + 14526.7\}/2.0$$

$$= 8460.39 \text{ W/m}^2 \text{ K}$$

6.12 For a power law velocity profile, $\frac{u}{u_e} = (\frac{y}{\delta})^{1/n}$,

$$\frac{\theta}{\delta} = \frac{n}{(n+1)(n+2)}, \quad \frac{\delta^*}{\delta} = \frac{1}{1+n}. \quad \text{For } n = 7, \quad \frac{\theta}{\delta} = \frac{7}{72} \qquad (1)$$

For a flat plate, $\frac{d\theta}{dx} = \frac{c_f}{2}$ can be written as

$$\frac{7}{72}\frac{dR_\delta}{dR_x} = \frac{c_f}{2} \quad \text{using (1).} \qquad (2)$$

For $c_f = 0.0456 R_\delta^{-1/4}$, it follows from (2), that,

$$R_\delta^{1/4} dR_\delta = \frac{0.0456}{2}\frac{72}{7} dR_x, \quad R_\delta = 0.37 R_x^{0.8}$$

or $\frac{\delta}{x} = 0.37 R_x^{-0.2}$, $\frac{\theta}{x} = 0.036 R_x^{-0.2}$,

$$c_f = 0.0456 \, [0.37 R_x^{0.8}]^{-1/4} = 0.059 R_x^{-0.2}$$

and $\bar{c}_f = \frac{1}{x}(\int_0^x c_f dx) = 0.074 R_x^{-0.20}$

6.13 From (6.33) and (6.35)

$$\frac{k_s}{k} = \exp[\kappa(\Delta u^+ - \Delta u_s^+)]$$

Here k_s and k are the heights of the equivalent sand grain and square-bar roughness elements. From Fig. 6.7 on page 167, $\Delta u^+ - \Delta u_s^+ = 3.25$. Thus, with $\kappa = 0.41$ and $k = 0.0005m$, $k_s = 3.79k = 0.0019m$ and

$$R_k = \frac{u_e}{\nu} k_s = 10^7 \times 0.0019 = 1.9 \times 10^4$$

At $x = 1m$, $\frac{x}{k_s} = 526.3$ and $R_x = 10^7$, $c_f = 6.7 \times 10^{-3}$ from Fig. 6.12.

At $x = 3m$, $R_x = 3 \times 10^7$, $\bar{c}_f = 6.5 \times 10^{-3}$ from Fig. 6.13.

6.14 a. Camouflage paint with $k_s = 1 \times 10^{-3}$cm

$$\frac{x}{k_s} = \frac{300}{1 \times 10^{-3}} = 3 \times 10^5, \qquad R_x = 3 \times 10^7, \qquad R_k = 100$$

From Figs. 6.12 and 6.13, c_f and $\bar{c}_f$ are, respectively,

$$c_f = 2.1 \times 10^{-3} \quad \text{and} \quad \bar{c}_f = 2.5 \times 10^{-3}$$

From the momentum integral equation for a flat plate $\bar{c}_f/2 = \theta/x$ we can write $x(\bar{c}_f/2) = 300\ (2.5 \times 10^{-3})/2 = 0.375$cm and $\delta = \frac{72}{7}\theta = 3.857$cm

b. For the case of iron with $k_s = 25 \times 10^{-3}$cm,

$$\frac{x}{k_s} = \frac{300}{25 \times 10^{-3}} = 1.2 \times 10^4, \qquad R_x = 3 \times 10^7, \qquad R_k = 2500$$

From Fig. 6.12 and 6.13, c_f and $\bar{c}_f$ are, respectively,

$$c_f = 3.5 \times 10^{-3} \quad \text{and} \quad \bar{c}_f = 4.2 \times 10^{-3}$$

The momentum thickness and boundary-layer thickness are

$$\theta = \bar{c}_f/2 \cdot x = 0.63\text{cm}, \quad \delta = \frac{72}{7}\theta = 6.48\text{cm}$$

6.15 For a flat plate flow, write the continuity and momentum equations as

$$\frac{\partial u}{\partial x} + \frac{\partial v}{\partial y} = 0, \quad u\frac{\partial u}{\partial x} + v\frac{\partial u}{\partial y} = \frac{1}{\rho}\frac{\partial \tau}{\partial y}$$

Multiply continuity by u and add the resulting expression to the momentum equation, then integrate the resulting equation wrt y.

$$\int_0^y \frac{\partial}{\partial x}\left(\frac{u^2}{u_e^2}\right)dy - \frac{u}{u_e}\int_0^y \frac{\partial}{\partial x}\left(\frac{u}{u_e}\right)dy = \frac{1}{u_e^2}\frac{\tau - \tau_w}{\rho} \tag{1}$$

For a power-law velocity profile, $\frac{u}{u_e} = \left(\frac{y}{\delta}\right)^{1/n} = \eta^{1/n} = g(\eta)$, $\eta = \frac{y}{\delta}$

then
$$\int_0^y \frac{\partial}{\partial x}\left(\frac{u^2}{u_e^2}\right)dy = -\frac{d\delta}{dx}\left(\eta g^2 - \int_0^\eta g^2 d\eta\right)$$
$$\int_0^y \frac{\partial}{\partial x}\left(\frac{u}{u_e}\right)dy = -\frac{d\delta}{dx}\left(\eta g - \int_0^\eta g\, d\eta\right) \tag{2}$$

Substitute (2) into (1),

$$\frac{\tau}{\tau_w} = 1 + \frac{2}{c_f}\left[\frac{d\delta}{dx}\left(\int_0^\eta g^2 d\eta - g\int_0^\eta g d\eta\right)\right] \quad (3)$$

and evaluate $d\delta/dx$ from $d\theta/dx = c_f/2$. Since δ/θ = constant for a power-law profile,

$$\frac{d\delta}{dx} = \frac{d\theta}{dx}\left(\frac{\theta}{\delta}\right)^{-1} = \frac{c_f}{2}\left(\frac{\theta}{\delta}\right)^{-1} \text{ or } \frac{2}{c_f}\frac{d\delta}{dx} = \left(\frac{\theta}{\delta}\right)^{-1} = \frac{(n+1)(n+2)}{n} \quad (4)$$

$$\text{then } \int_0^\eta g^2 d\eta = \int_0^\eta \eta^{2/n} d\eta = \frac{n}{n+2}\eta^{(n+2)/n}$$

$$g\int_0^\eta g d\eta = \eta^{1/n}\int_0^\eta \eta^{1/n} d\eta = \frac{n}{n+1}\eta^{(n+2)/n} \quad (5)$$

Then substitute (4) and (5) into (3),

$$\frac{\tau}{\tau_w} = 1 + \frac{(n+1)(n+2)}{n}\left[\frac{n}{n+2}\eta^{(n+2)/n} - \frac{n}{n+1}\eta^{(n+2)/n}\right]$$

$$= 1 - \eta^{1+2/n} = 1 - \left(\frac{y}{\delta}\right)^{1+2/n}$$

6.16 For power-law velocity and temperature profiles,

$$\frac{u}{u_e} = \left(\frac{y}{\delta}\right)^{1/n}, \quad \frac{T_w - T}{T_w - T_e} = \left(\frac{y}{\delta_t}\right)^{1/n}$$

$$\frac{\partial u}{\partial y} = \frac{u_e}{n}\frac{1}{\delta}\left(\frac{y}{\delta}\right)^{1/n-1}, \quad -\frac{\partial T}{\partial y} = \frac{T_w - T_e}{n}\left(\frac{y}{\delta_t}\right)^{1/n-1}\frac{1}{\delta_t}$$

$$\text{With } Pr_t = 1, \quad \frac{\varepsilon_m}{\varepsilon_h} = 1.0 = \frac{(\tau/\rho)/(\partial u/\partial y)}{\overline{-T'v'}/(\partial T/\partial y)}$$

$$\therefore\ \overline{-T'v'} = \frac{\tau}{\rho}\frac{\partial T}{\partial y}\left(\frac{\partial u}{\partial y}\right)^{-1} = -\frac{\tau}{\rho}\frac{T_w - T_e}{u_e}\left(\frac{\delta}{\delta_t}\right)^{1/n}$$

$$\dot{q}_w \simeq \dot{q} = \rho c_p \overline{T'v'} = \rho c_p \frac{\tau_w}{\rho}\frac{T_w - T_e}{u_e}\left(\frac{\delta}{\delta_t}\right)^{1/n}$$

$$\text{Then } St = \frac{\dot{q}_w}{\rho c_p u_e (T_w - T_e)} = \frac{c_p \tau_w (T_w - T_e)}{\rho c_p u_e^2 (T_w - T_e)}\left(\frac{\delta}{\delta_t}\right)^{1/n} = \frac{c_f}{2}\left(\frac{\delta_t}{\delta}\right)^{-1/n}$$

6.17 a. From (6.79), for $x_0 = 0$,

$$StPr^{0.4}\left(\frac{T_w}{T_e}\right)^{0.4} = 0.0296\ R_x^{-0.20} = 0.0296\ R_L^{-0.20}\left(\frac{x}{L}\right)^{-0.20}$$

When $R_L = 9 \times 10^5$, $Pr = 0.7$, $T_w/T_e = 1.1$ and $L = 3m$

$$St = \frac{0.0296\ (9 \times 10^5)^{-0.20}}{(0.7)^{0.4}(1.1)^{0.4}}\left(\frac{x}{3}\right)^{-0.20} = 2.64 \times 10^{-3} x^{-0.20} \quad (a)$$

b. For $x_0 > 0$, (6.79) becomes

$$StPr^{0.4}\left(\frac{T_w}{T_e}\right)^{0.4} = 0.0296\ R_x^{-0.20}\left[1 - \left(\frac{x_0}{x}\right)^{9/10}\right]^{-1/9}$$

When $R_L = 9 \times 10^5$, $Pr = 0.7$, $T_w/T_e = 1.1$, $L = 3m$, $x_0 = 1m$,

$$St = 2.64 \times 10^{-3}\ x^{-0.20}\left[1 - \left(\frac{1}{x}\right)^{9/10}\right]^{-1/9} \qquad \text{(b)}$$

The variations of St with x for (a) and (b) are shown in the figure.

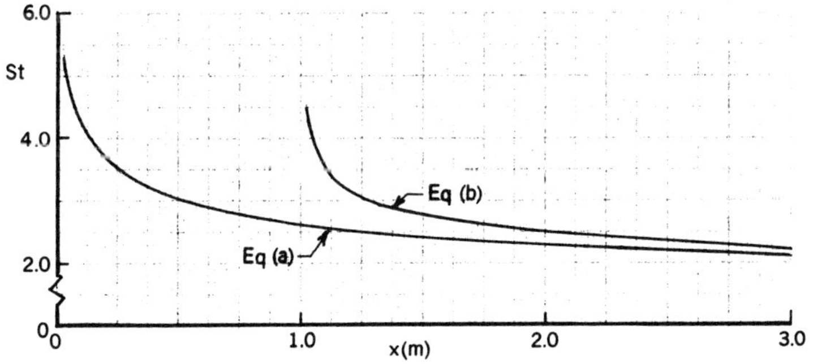

Discussion: Since $(T_w/T_e)^{0.4} > 1.0$, if $T_w/T_e > 1.0$ and vice versa, St decreases as the wall is heated and increases as the wall is cooled.

6.18 To calculate the heat flux, $\dot{q}_w$, we use (6.82). For the present case

$$\frac{dT_w}{dx} = \frac{d}{dx}(A + Bx) = B, \quad N = 1, \quad x_{01} = 0$$

$$T_w(x_{01}^+) - T_w(x_{01}^-) = T_w(0^+) - T_w(0^-) = A$$

$$\hat{h}(x_{01},x) = \rho c_p u_e St_T\left[1 - \left(\frac{0}{x}\right)^{9/10}\right]^{-1/9} = \rho c_p u_e St_T$$

so $$\dot{q}_w(x) = \rho u_e c_p St_T(x)\left\{A + B\int_0^x \left[1 - \left(\frac{\xi}{x}\right)^{9/10}\right]^{-1/9} d\xi\right\}$$

$$= \rho u_e c_p St_T(x)\left\{A + Bx\int_0^1 \left[1 - \eta^{9/10}\right]^{-1/9} d\eta\right\}$$

$$\int_0^1 (1 - \eta^{9/10})^{-1/9} d\eta = \frac{10}{9}\int_0^1 \xi^{-1/9}(1 - \xi)^{1/9} d\xi = \frac{10}{9} B_1\left(\frac{8}{9}, \frac{10}{9}\right)$$

Here $B_r(m,n) = \int_0^r \xi^{m-1}(1 - \xi)^{n-1} d\xi$

$$\therefore\ \dot{q}_w(x) = \rho u_e c_p St_T(x)\left[A + \frac{10B}{9} xB_1(8/9, 10/9)\right]$$

6.19 The heat flux on a flat plate subject to a series of discontinuous wall temperature is

$$\dot{q}_w = \sum_{n=1}^{N} \hat{h}(x_{on},x)[T_w(x_{on}^+) - T_w(x_{on}^-)]$$

For the present case, N = 3 and

$$n = 1, \quad \hat{h}(0,x) = \rho u_e c_p St_T(x), \quad T_w(0^+) - T_w(0^-) = T_{w_1} - T_e$$

$$n = 2, \quad \hat{h}(x_1,x) = \rho u_e c_p St_T(x) \, [1 - (\frac{x_1}{x})^{9/10}]^{-1/9}$$

$$T_w(x_1^+) - T_w(x_1^-) = T_{w_2} - T_{w_1}$$

$$n = 3, \quad \hat{h}(x_2,x) = \rho u_e c_p St_T(x) \, [1 - (\frac{x_2}{x})^{9/10}]^{-1/9}$$

$$T_w(x_2^+) - T_w(x_2^-) = T_{w_3} - T_{w_2}$$

so
$$\dot{q}_w(x) = \rho u_e c_p St_T(x) \, \{(T_{w_1} - T_e) + (T_{w_2} - T_{w_1})[1 - (\frac{x_1}{x})^{9/10}]^{-1/9}$$
$$+ (T_{w_3} - T_{w_2})[1 - (\frac{x_2}{x})^{9/10}]^{-1/9}\}$$

6.20 Use the expression for $\dot{q}_w$ given in Problem 6.19 and with $T_{w_1} = 50°C$, $T_{w_2} = 60°C$, $T_{w_3} = 80°C$, $T_e = 20°C$, $x_1 = 1m$, $x_2 = 1.5m$, $x = 4m$, $\frac{u_e}{\nu} = 5 \times 10^5/m$ and $k = 0.0279$ W/mK,

$$\dot{q}_w(x) = 5 \times 10^5 \times 0.0279 \times 0.0296 \times (2 \times 10^6)^{-0.20}\{(50 - 20)$$
$$+ (60 - 50)[1 - (\frac{1}{4})^{9/10}]^{-1/9} + (80 - 60)[1 - (\frac{1.5}{4})^{9/10}]^{-1/9}\}$$
$$= 1397 \text{ W/m}^2$$

6.21 The wall temperature is calculated from Eq. (6.87), which is

$$T_w(x) - T_e = \frac{33.61 \, \dot{q}_w Pr^{0.4} R_x^{0.2}}{\rho c_p u_e} \frac{\beta_r(1/9, 10/9)}{\beta_1(1/9, 10/9)}$$

$$= \frac{33.61 \, \dot{q}_w Pr^{-0.6} R_x^{0.2}}{u_e/\nu \; k} \frac{\beta_r(1/9, 10/9)}{\beta_1(1/9, 10/9)}$$

For $\frac{u_e}{\nu} = 5 \times 10^5/m$, $R_x = 5 \times \frac{u_e}{\nu} = 25 \times 10^5$, $k = 0.02535$, $Pr = 0.7$, $x_o = 2m$, $x = 5m$, $r = 1 - (x_o/x)^{9/10} = 0.5616$, and $\dot{q}_w$ in W/m^2

$$T_w(x) - T_e = \frac{33.61 \times (0.7)^{-0.6} \times (2.5\times10^6)^{0.2}}{5 \times 10^5 \times 0.02535} \times 0.946 \times \dot{q}_w$$

$$= 0.0591\ \dot{q}_w$$

6.22 First, with $\eta = \xi/x$ and $\eta_0 = x_0/x$, consider the integrand,

$$\int_{x_0}^{x} [1 - (\tfrac{\xi}{x})^{9/10}]^{-8/9} d\xi = x \int_{\eta_0}^{1} (1 - \eta^{9/10})^{-8/9} d\eta$$

Let $z = 1 - \eta^{9/10}$, $dz = -\frac{9}{10}(1 - z)^{-1/9} d\eta$ and with $\beta_r(m,n)$ denoting incomplete beta function and $r = 1.0 - (x_0/x)^{9/10}$,

$$x \int_{\eta_0}^{1} [1 - \eta^{9/10}]^{-8/9} d\eta = x \int_0^r z^{-8/9}(1 - z)^{1/9}\, \frac{10}{9}\, dz = \frac{10x}{9}\, \beta_r(\tfrac{1}{9}, \tfrac{10}{9})$$

Then from Eq. (6.86), $T_w - T_e = \dfrac{3.42\dot{q}_w}{Pr^{0.6} R_x^{0.8} k} \displaystyle\int_{x_0}^{x} [1 - (\tfrac{\xi}{x})^{9/10}]^{-8/9} d\xi$

$$= \frac{3.42 Pr^{0.4} R_x^{0.2} \dot{q}_w}{(\mu c_p/k)(u_e x \rho/\mu)k}\, \frac{10x}{9}\, \beta_1(\tfrac{1}{9}, \tfrac{10}{9})\, \frac{\beta_r(1/9,10/9)}{\beta_1(1/9,10/9)}$$

$$= \frac{3.42 \times 10}{9} \times 8.844\, \frac{\dot{q}_w Pr^{0.4} R_x^{0.2}}{\rho c_p u_e}\, \frac{\beta_r(1/9,10/9)}{\beta_1(1/9,10/9)}$$

$$= \frac{33.61 \dot{q}_w Pr^{0.4} R_x^{0.2}}{\rho c_p u_e}\, \frac{\beta_r(1/9,10/9)}{\beta_1(1/9,10/9)}$$ which is Eq. (6.87) with

$\beta_1(1/9,10/9) = 8.844$.

6.23 With the law of the wall, $u = u_\tau f\,(\frac{y u_\tau}{\nu})$ and with $f' = \frac{df}{dy^+}$, $y^+ = \frac{y u_\tau}{\nu}$

$$\frac{\partial u}{\partial x} = \frac{du_\tau}{dx} f + f' u_\tau \frac{y}{\nu} \frac{du_\tau}{dx} = \frac{du_\tau}{dx} (f + \frac{y u_\tau}{\nu} f')$$

From the continuity equation,

$$v = -\int_0^y \frac{\partial u}{\partial x} dy = -\frac{\nu}{u_\tau} \frac{du_\tau}{dx} \int_0^{y^+} (f + y^+ f') dy^+$$

$$= -\frac{\nu}{u_\tau} \frac{du_\tau}{dx} \{ \int_0^{y^+} f dy^+ + f y^+ - \int_0^{y^+} f dy^+ \} = -y \frac{u}{u_\tau} \frac{du_\tau}{dx}$$

6.24 $V_E \equiv \dfrac{d}{dx} \displaystyle\int_0^\delta u dy = \frac{d}{dx} [\int_0^\delta u_e dy - \int_0^\delta (u_e - u) dy] = \frac{d}{dx} [u_e(\delta - \delta^*)]$

$$\therefore \frac{V_E}{u_e} = \frac{1}{u_e} \frac{d}{dx} [u_e(\delta - \delta^*)]$$

6.25 $$V_E \equiv \frac{1}{r_\delta}\frac{d}{dx}\int_0^\delta rudy = \frac{1}{r_\delta}\frac{d}{dx}\int_0^\delta r(u_e - u_e + u)dy$$

$$= \frac{1}{r_\delta}\frac{d}{dx}\left[\int_0^\delta u_e rdy - u_e \int_0^\delta r(1 - \frac{u}{u_e})dy\right.$$

where $r = r_0 + y\cos\phi(x)$, $\delta^* \equiv \int_0^\delta r(1 - \frac{u}{u_e})dy$

then $$V_E = \frac{1}{r_\delta}\frac{d}{dx}\{u_e\delta[r_0 + \frac{\delta}{2}\cos\phi(x)] - \delta^* u_e\}$$

$$= \frac{1}{r_\delta}\frac{d}{dx}\{u_e\delta[\frac{r_0}{2} + \frac{r_0 + \delta\cos\phi(x)}{2}] - \delta^* u_e\};$$

since $r_\delta = r_0 + \delta\cos\phi(x)$

$$V_E = \frac{1}{r_\delta}\frac{d}{dx}[u_e(\delta\frac{r_0 + r_\delta}{2} - \delta^*)]$$

6.26 a. $\eta^* = \frac{T_{aw} - T_e}{T_a - T_e} = \frac{300 - 275}{325 - 275} = 0.5$; $\rho_e = 1.295\ kg/m^3$ and

$\rho_c = 1.088\ kg/m^3$ for $T_e = 275K$ and $T_a = 325K$

$$F = \frac{\rho_c u_c}{\rho_e u_e} = \frac{1.088 \times 25}{1.295 \times 50} = 0.42$$

Assume $x/h \geq 20$ but $x/h \leq 150$. Then from Eq. (6.131a)

$\eta^* = 5.44F^{0.4}(\frac{x}{h})^{-0.58}$. For $\eta^* = 0.5$, $F = 0.42$, $\frac{x}{h} = 33.7 \geq 20.0$,

$$h = \frac{1}{33.7} = 0.0297m = 2.97\ cm$$

b. $\nu = 12.6 \times 10^{-6}\ m^2/s$ for $T_e = 275K$;

$$R_{x_0} = \frac{u_e x_0}{\nu} = \frac{50.0 \times 2.0}{12.6 \times 10^{-6}} = 7.94 \times 10^6$$

For a turbulent flow starting at the leading edge,

$\frac{\delta}{x_0} = 0.37(R_{x_0})^{-0.20}$, $\delta = 2.0 \times 0.37\ (7.94 \times 10^6)^{-0.20} = 0.0308\ m$

$$\frac{\delta}{y_c} = \frac{3.08}{2.97} = 1.04$$

Therefore the calculated δ/y_c is much smaller than 2.0 used in Fig. 6.28. However, the c_f for $\delta/y_c = 2.0$ should not be too different from its actual value since c_f is a weak function of δ/y_c, as demonstrated in Figs. 6.28 and 6.30 for which the values of δ/y_c are 2.0 and 0.95, respectively.

6.27 $\nu = 14.2 \times 10^{-6}$ and $k = 0.0252$ W/m K for $T_f = \frac{1}{2}(T_w + T_e) = 287.5$K

since $L = y_c = 0.0297$ m, $\xi = (x - x_0)/y_c$,

$$R_L = \frac{u_e y_c}{\nu} = \frac{50.0 \times 0.0297}{14.2 \times 10^{-6}} = 1.046 \times 10^5, \quad T_c = 325\text{K}, \quad T_e = 275\text{K}$$

then
$$\dot{q}_w = \frac{g_w'(T_c - T_e)R_L^{1/2}k}{\xi^{1/2}L} = \frac{g_w'}{\xi^{1/2}} \times 50 \times (1.046 \times 10^5)^{1/2} \times \frac{0.0252}{0.0297}$$

$$= 1.37 \times 10^4 \frac{g_w'}{\xi^{1/2}}$$

$x-x_0$	ξ	g_w'	$\dot{q}_w$	$x-x_0$	ξ	g_w'	$\dot{q}_w$
0	0	0	-	0.60	20.3	-0.38	-1155.5
0.25	8.5	0.01	-47.0	0.70	23.7	-0.48	-1350.8
0.30	10.2	-0.02	-85.8	0.80	27.1	-0.60	-1579.0
0.40	13.56	-0.11	-409.2	0.90	30.5	-0.69	-1711.7
0.50	17.0	-0.24	-797.5	1.00	33.9	-0.78	-1835.3

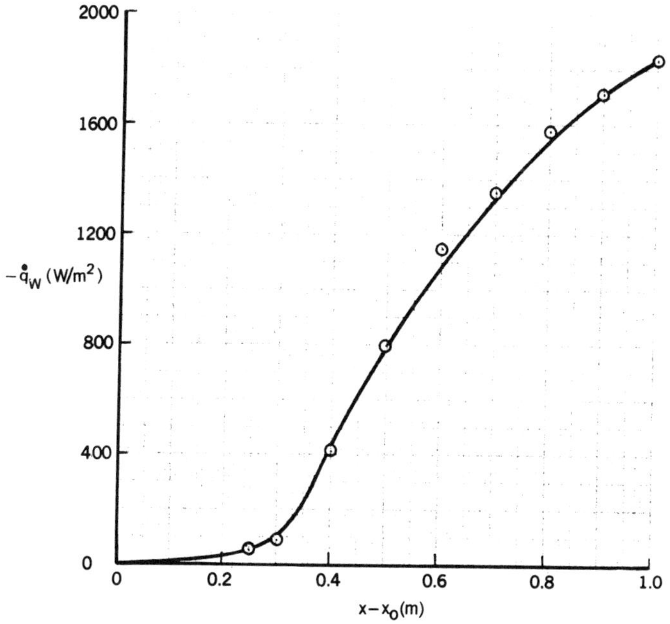

6.28 Consider a flow of air which is initially laminar and becomes turbulent around a two-dimensional ellipse of thickness ratio 4, with $T_\infty = 15°$C and $T_w = 25°$C, $R_{2a} = 10^7$, Pr = 0.72. The location of transition is obtained from Eq. (6.111) where $R_{x_{tr}}$ is the Reynolds number based on the surface distance. Laminar flow calculations are performed with the computer program given in Problem 4.28. For turbulent flow, Eqs. (6.128), (6.129a) and (6.129b) are used for momentum transfer and Eq. (6.130) for heat transfer. A computer program has been written to integrate Eqs. (6.128) to (6.130) in the code of Problem 4.28 and is given in the diskette. Note that Eq. (6.130) should be corrected to

$$St = \frac{\dot{q}_w}{\rho c_p u_e (T_w - T_e)} = \frac{Pr^{-0.6} R_L^{-0.2} (T_w - T_e)^{0.25}}{\left[\int_{x_{tr}}^{x^*} u_e^* (T_w - T_e)^{1.25} dx^*\right]^{0.2} + C_1}$$

The results obtained with this code (see the diskette) for $c_f/2\ \sqrt{R_x}$ vs x/a and $Nu_x/\sqrt{R_x}$ vs (x/a) are shown below.

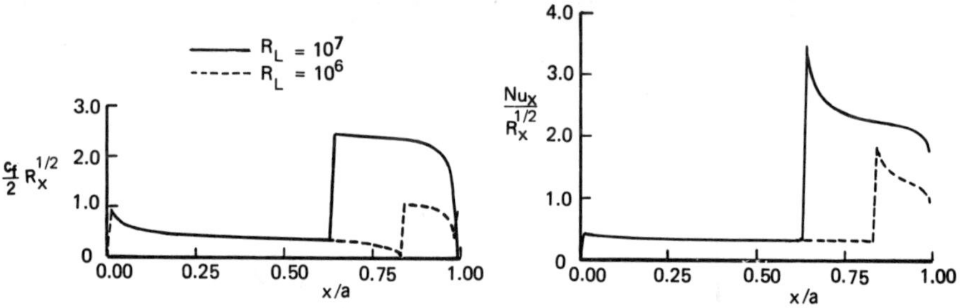

6.29 This problem is similar to Problem 6.28 and is for the NACA 2412 airfoil whose coordinates as well as the external velocity distribution are given. The upper and lower surfaces are considered separately with the computer program of Problem 6.28. The coordinates x/c, y/c and the external velocity distribution u_e/u_∞ as a function of x/c are input for both upper and lower surfaces from different data files. However, it is important to note that 17 points for each surface are insufficient for accurate calculations and should be increased to, say, 34 by linear interpolation. The computer program is given in the diskette. The derivative of $u_e(x)$ wrt x should be obtained from a three-point Lagrange interpolation scheme to avoid possible oscillations of the solutions. Note that the transition location on the upper surface is at $x/c = 0.7$, and at $x/c = 0.65$ on the lower surface. The surface temperature of the airfoil is kept constant at a value of 550°C up to the point of transition, and is then cooled down to 450°C at the trailing edge.

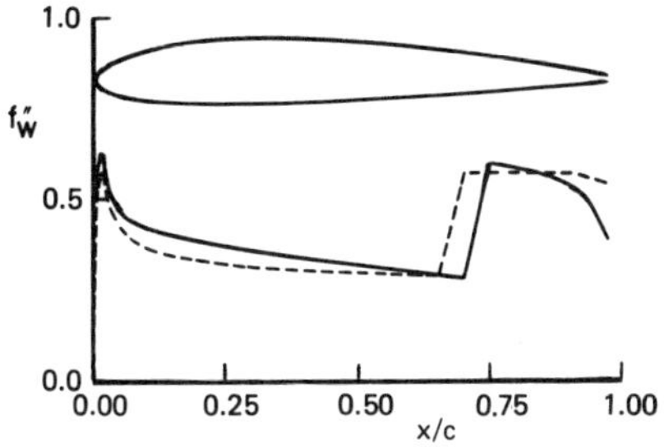

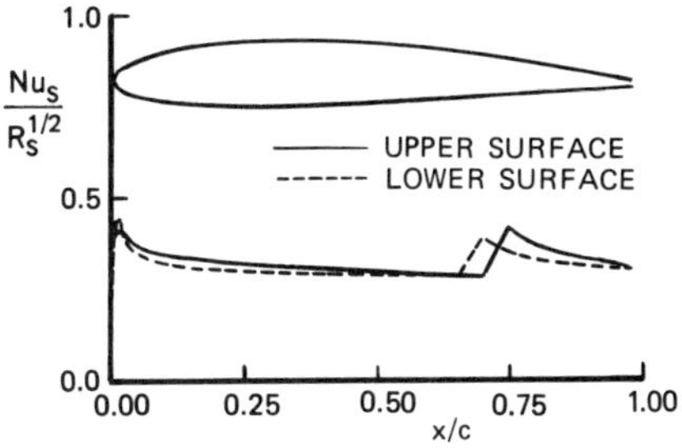

6.30 Laminar flow calculations are performed with Thwaites' method and those for turbulent flow with Head's method. In the axisymmetric case, the values of x(I) are replaced by the surface distance, and the values of $r_0(x)/L$ (or $y(x)/L$) are input to the computer program. The necessary changes to the computer program of Appendix D for axisymmetric flow are as follows:

a. Modify common statement to include the arrays R(41) and DRDX(41) where R(I) is the array containing r_0/L (x/L) and DRDX(I) is the array containing $[d(r_0/L)]/[d(x/L)]$.

b. Use the three-point Lagrange interpolation formula to calculate $[d(r_0/L)]/[d(x/L)]$:

```
DRDX(1) = RGNG[X(1),X(2),X(3),R(1),R(2),R(3),X(1)]
DRDX(NXT) = RGNG[X(NXT-2),X(NXT-1),X(NXT),R(NXT-2),R(NXT-1),
            R(NXT),X(NXT)]
```

c. Modify DO loop 10 as:

```
DO 10 I = 2, NXTM1,
DUEDX = ...,
DRDX(I) = RGNG[X(I-1),X(I),X(I+1),R(I-1),R(I),R(I+1),X(I)]
```

d. Modify the expression S(1) as:

```
S(1) = UE(1)IT(1)*R(1)*HOFH1(-H(1))
```

e. In subroutine STNDRD, modify;

```
C(1) = -(HB+2.0)*B(1)/UI*UP+CF02-B(1)/R(1)*UP*DRDX(I)
H1 = B(2)/B(1)/UI/R(1)
C(2) = UI*0.0306/(H1-3.0)**0.6169*R(I)
```

Note that X is the surface distance and the derivatives must be evaluated with respect to their surface coordinates. See the diskette for the computer program.

Chapter 7

Uncoupled Turbulent Duct Flows

7.1 Since $\tau_w = \rho u_\tau^2 = \mu \left(\frac{du}{dy}\right)_{y=0} = -\mu \left(\frac{du}{dr}\right)_{r=r_0}$ or $\left(\frac{du}{dr}\right)_{r=r_0} = -\frac{u_\tau^2}{\nu}$ (1)

Eq. (7.6) can be written at $r = r_0$ as

$$r_0^K \left(\frac{du}{dr}\right)_{r=r_0} = \frac{r_0^{2K}}{(K+1)} \frac{1}{\nu} \frac{d\bar{P}}{dx} \tag{2}$$

From Eqs. (1) and (2), it follows that

$$-r_0^K \frac{u_\tau^2}{\nu} = \frac{r_0^{2K}}{K+1} \frac{1}{\nu} \frac{d\bar{P}}{dx} = \left(\frac{r_0}{r}\right)^{2K} \frac{r^{2K}}{K+1} \frac{1}{\nu} \frac{d\bar{P}}{dx} = \left(\frac{r_0}{r}\right)^{2K} b \frac{du}{dr}$$

$$= -\left(\frac{r_0}{r}\right)^{2K} b \frac{du}{dy} \tag{3}$$

With $b = r^K(1 + \varepsilon_m^+)$, Eq. (3) can be written in the form of Eq. (7.7)

$$\left(\frac{r_0}{r}\right)^{2K} r^K(1 + \varepsilon_m^+) \frac{du}{dy} = r_0^K \frac{u_\tau^2}{\nu} \quad \text{or} \quad (1 + \varepsilon_m^+) \frac{du}{dy} = \left(\frac{r}{r_0}\right)^K \frac{u_\tau^2}{\nu}$$

7.2 By definition $f = \dfrac{\tau_w}{(1/8)\rho u_m^2} = \dfrac{u_\tau^2}{(1/8)u_m^2}$ or $u_\tau = u_m \sqrt{f/8}$

If $u_m = u_{max} - 4.07u_\tau$

Then $u_{max} = u_m + 4.07u_\tau = u_m[1 + 4.07 \sqrt{f/8}] = u_m(1 + 1.44 \sqrt{f})$

7.3 In pipe flow Π is negligibly small; the velocity at $y = r_0$ can then be expressed as

$$u_{max}^+ = \frac{1}{\kappa} \ln \frac{r_0 u_\tau}{\nu} + c = \frac{1}{\kappa} \ln \left(\frac{r_0 u_m}{\nu} \sqrt{f/8}\right) + c \tag{1}$$

where u_{max}^+ is related to u_m^+ by $u_{max}^+ = u_m^+(1 + 1.44 \sqrt{f})$ (2)

Combining (1) and (2), $u_m^+(1 + 1.44 \sqrt{f}) = \frac{1}{\kappa} \ln \left[\frac{r_0 u_m}{\nu} \sqrt{f/8}\right] + c$ or

$$\frac{1}{\sqrt{f/8}}(1 + 1.44\sqrt{f}) = \frac{1}{\kappa}[\ln \sqrt{f} R_d - 0.5 \ln 8] + c$$

or $$\frac{1}{\sqrt{f}} = \frac{1}{\sqrt{8}}\frac{1}{\kappa}\ln(\sqrt{f} R_d) + [\frac{1}{\sqrt{8}}(-\frac{0.5}{\kappa}\ln 8 + c) - 1.44]$$

Let $A = \frac{1}{\sqrt{8}}\frac{1}{\kappa}$, $B = \frac{1}{\sqrt{8}}(c - \frac{0.5}{\kappa}\ln 8) - 1.44$ then $\frac{1}{\sqrt{f}} = A \ln R_d\sqrt{f} + B$

which is in the form of Prandtl's friction law for smooth pipes. For $\kappa = 0.41$ and $c = 5.2$, $A = 0.86$ and $B = -0.5$. For better agreement with the experimental data, set $A = 0.87$ and $B = -0.8$, which is the Prandtl law given by (7.16).

7.4 Again, since Π is negligibly small in channel flow, the entire velocity profile can be represented by $u^+ = \frac{1}{\kappa}\ln\frac{yu_\tau}{\nu} + c$ and be used to find u_{max} and u_m.

$$u^+_{max} = \frac{1}{\kappa}\ln\frac{hu_\tau}{\nu} + c, \quad u^+_m = \frac{1}{\kappa}(\ln\frac{hu_\tau}{\nu} - 1.0) + c \tag{1}$$

$$\therefore \quad u^+_m = u^+_{max} - \frac{1}{\kappa} \tag{2}$$

From the second equation in (1), we can write,

$$\frac{1}{\sqrt{f/8}} = \frac{1}{\kappa}[\ln \sqrt{f/8}\, R - 1.0] + c$$

$$\frac{1}{\sqrt{f}} = \frac{1}{\sqrt{8}}\frac{1}{\kappa}[\ln \sqrt{f} R - 0.5 \ln 8 - 1.0] + \frac{c}{\sqrt{8}}$$

$$= \frac{1}{\sqrt{8}}\frac{1}{\kappa}[\ln \sqrt{f} R] + \frac{c}{\sqrt{8}} - \frac{1}{\sqrt{8}\,\kappa}(0.5 \ln 8 + 1.0)$$

Let $\frac{1}{\kappa}\frac{1}{\sqrt{8}} = A$, and $\frac{1}{\sqrt{8}}[c - (0.5 \ln 8 + 1.0)/\kappa] = B$

Then $\frac{1}{\sqrt{f}} = A \ln R\sqrt{f} + B$

is in the form of Prandtl's friction law for turbulent flow in a smooth channel. With $\kappa = 0.41$ and $c = 5.2$, $A = 0.86$ and $B = 0.08$. For better agreement with experimental data, set $A = 0.87$, $B = -0.41$, we have Prandtl's friction law, expressed as

$$\frac{1}{\sqrt{f}} = 0.87 \ln R\sqrt{f} - 0.41$$

7.5 With the matrix elimination procedure, we first construct a matrix whose columns give the mass, length, time and temperature dimensions of the variables in each row

	M	L	T	θ
$\hat{h}$	1	0	-3	-1
u_o	0	1	-1	0
ρ	1	-3	0	0
μ	1	-1	-1	0
c_p	0	2	-2	-1
k	1	1	-3	-1
d	0	1	0	0

We first eliminate M, by dividing the relevant variables by ρ, then θ by dividing by k/ρ.

	L	T
$\hat{h}/k$	-1	0
u_o	1	-1
μ/ρ	2	-1
$c_p\rho/k$	-2	1
d	1	0

Next we eliminate T and L by forming $\hat{h}d/k$, $u_od/(\mu/\rho)$ and $\mu c_p/k$, or

$$\frac{\hat{h}d}{k} = f\left(\frac{u_o d}{\mu/\rho}, \frac{\mu c_p}{k}\right)$$

The first group $\hat{h}d/k$ is the Nusselt number by definition, the second is the Reynolds number, $R_d = u_od/(\mu/\rho)$ and the last one is the Prandtl number, $Pr = \mu c_p/k$. Therefore, $Nu = f(R_d, Pr)$.

7.6 See the diskette for the computer program. Computed results for uniform heat flux at two Reynolds numbers are given below.

Pr	$Nu(R_d = 10^4)$	$Nu(R_d = 10^5)$
0.02	6.78	15.61
0.72	29.00	172.69
14.3	106.07	831.66

7.7 See the diskette for the computer program. Computed results for the uniform wall temperature at two Reynolds numbers are given below.

Pr	$Nu(R_d = 10^4)$	$Nu(R_d = 10^5)$
0.02	5.39	13.80
0.72	27.89	169.39
14.3	105.75	830.59

7.8 The use of Eqs. (7.23), (7.24), (7.25) and (7.26) for evaluating the Nusselt number yield:

(a) $Pr = 0.7$, $R_d = 10{,}000$, $f = 0.3164/R_d^{0.25} = 0.03164$

	Eq. (7.23)	Eq. (7.24)	Eq. (7.25)	Eq. (7.26)
Nu	39.95	36.4	32.7	53.3

(b) Pr = 0.2, R_d = 10,000

	Eq. (7.23)	Eq. (7.24)	Eq. (7.26)
Nu	14.6	15.7	24.4

(c) Pr = 100, R_d = 10,000

	Eq. (7.23)	Eq. (7.24)	Eq. (7.26)
Nu	140.6	3045.0	221.9

(d) Pr = 0.7, $R_d = 10^6$, $f = 0.3164/R_d^{0.25} = 0.01$

	Eq. (7.23)	Eq. (7.24)	Eq. (7.26)
Nu	390.0	1515.2	401.4

7.9 $T_m = 0.5(T_e + T_w) = 0.5(50 + 110.0) = 80°C$. For liquid mercury with $T_m = 80°C$, $\rho = 13434.6\ kg/m^3$, $\nu = 0.097 \times 10^{-6}\ m^2/s$, $\kappa = 5.42 \times 10^{-6}\ m^2/s$, Pr = 0.018. Mass flux $\dot{m} = \rho u_m A = 13434.58 \times \pi \times (0.01)^2 u_m$ = 2.0.

$$\therefore u_m = 0.474\ m\ s^{-1} \quad \text{and} \quad R_d = \frac{u_m d}{\nu} = \frac{0.474 \times 0.02}{0.097 \times 10^{-6}} = 0.98 \times 10^5$$

For a low Prandtl number fluid, we calculate Nu from the formula proposed by Sleicher and Rouse, which is: $Nu = 6.3 + 0.0167\ R_d^{0.85} Pr^{0.93}$ $= 6.3 + 0.0167(0.98 \times 10^5)^{0.85}(0.018)^{0.93} = 13.24$. The total heat flux is

$$Q = \int_0^L \dot{q}_w p dx = \pi d L \dot{q}_w = \pi d L\ (T_w - T_m)\ Nu\ \frac{k}{d} = (T_w - T_m)\ Nu \pi k L$$

$$= c_p(T_o - T_i)\rho u_m \frac{\pi}{4} d^2$$

$$\therefore L = \frac{\rho c_p}{k} \frac{T_o - T_i}{T_w - T_m} \frac{u_m d^2 Nu^{-1}}{4} = \{54.2 \times 10^{-7}\}^{-1} \frac{45}{15} \frac{0.474}{4} \times (0.02)^2$$

$$\times (13.24)^{-1} = 1.98\ m/s$$

7.10

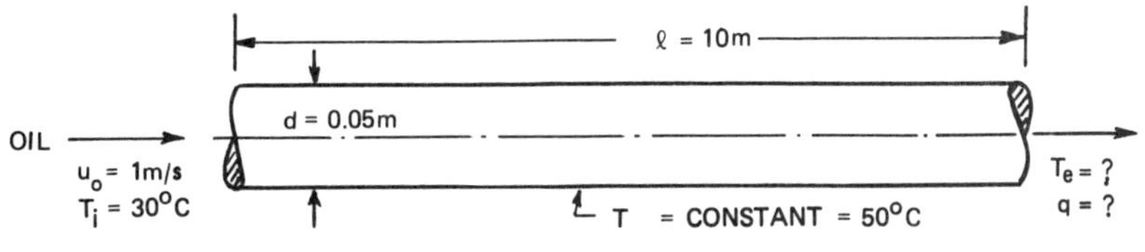

Since $T_w - T_b = 10°C$ and $T_w = 50°C$, the bulk temperature of the oil is 40°C. Evaluating its properties at $T_b = 40°$, we get

$\rho = 876.05\ kg/m^3$, $c_p = 1.964\ kj/kg\ K$,

$\nu = 0.00024\ m^2/s^1$, $k = 0.144\ W/m\ K$, Pr = 2870 so that

$$R_d = \frac{u_o d}{\nu} = \frac{(1)(0.05)}{0.00024} = 208$$

and the flow is laminar. The variation of Nusselt number with x obtained from Fig. 5.4 is summarized below.

$\hat{x}\ 10^4$	x	$\overline{Nu}$	$\hat{x}\ 10^4$	x	$\overline{Nu}$
1	0.746	30	7	5.22	14.5
2	1.49	23	10	7.46	13.0
3	2.24	20	13	9.69	12.0
4	2.98	18	14	10.44	11.5
5	3.73	16.5			

The total heat $\dot{Q}$ transferred to oil is

$$\dot{Q} = \int_0^{x=10m} h(x)\pi d(T_w - T_b)dx = \pi d(T_w - T_b) \int_0^{\ell} h(x)dx \qquad (1)$$

Since $Nu(x) = \frac{h(x)d}{k}$ $\quad h(x) = \frac{Nu(x)k}{d}$

Eq. (1) becomes $\dot{Q} = \pi(T_w - T_b)k \int_0^{\ell} Nu(x)dx$

and, with $(\Delta x)_i$ denoting the length of the ith segment along the pipe, it can be approximated as

$$\dot{Q} = \pi(T_w - T_b)k \sum_i Nu_i(\Delta x)_i = (\pi)(10)(0.144)(166.3) = 752.3 \text{ W}$$

The exit temperature T_e is found from $\dot{m}c_p(T_e - T_i) = \dot{Q}$

where $\dot{m} = \rho u \frac{\pi d^2}{4} = 1.72$ kg/s so that

$$T_e = \frac{\dot{Q}}{\dot{m}c_p} + T_i = \frac{752.3}{(1.72)(1964)} + 30 = 30 + 0.22 = 30.22°C$$

7.11

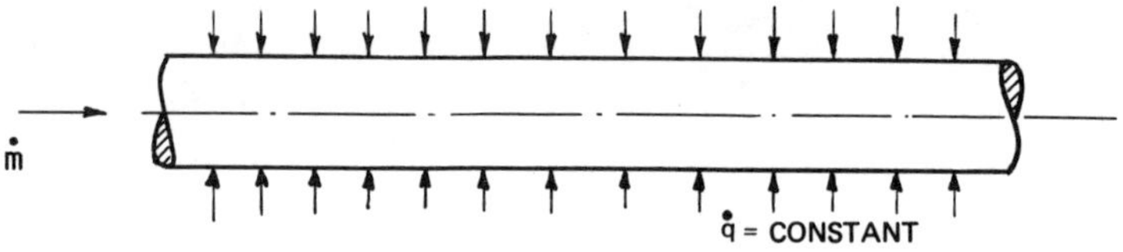

$\dot{m} = 1$ kg/s, $T = 20°C$, $d = 5 \times 10^{-2}$m, $P = 15$m, $T_w = 50°C$, $T_w - T_m = 10°C$. Evaluate properties of water at $20°C = 273 + 20 = 293$ K

$\rho = 1000$ kg/m^3, $c_p = 4.1818$ kJ/kg K, $\nu = 1.006 \times 10^{-6}$ m^2/s,

$k = 0.597$ W/mK, $Pr = 7.02$

To find the Reynolds number R_{r_o} and R_d

$$\bar{u} = \frac{4\dot{m}}{\rho \cdot \pi d^2} = \frac{(4)(1)}{(10^3)(\pi)(25 \times 10^{-4})}, \quad \bar{u} = \frac{40}{(25)(\pi)} = 0.5 \text{ m/s}$$

$$R_{r_o} = \frac{(2.5 \times 10^{-2})(0.5)}{1.006 \times 10^{-6}} = 12425, \qquad R_d \simeq 25,000$$

			$\overline{Nu}_i$	$h_i = \frac{Nu_i k}{d} =$
$\ell = 15m$	$\hat{x} = 6.84 \times 10^{-3}$	$\frac{x}{d} = 300$	174.5	2083.5 W/m^2 K
$\ell = 10m$	$\hat{x} = 4.56 \times 10^{-3}$	$\frac{x}{d} = 200$	174.5	2083.5 W/m^2 K
$\ell = 5m$	$\hat{x} = 2.28 \times 10^{-3}$	$\frac{x}{d} = 100$	174.5	2083.5 W/m^2 K
$\ell = 2.5m$	$\hat{x} = 1.14 \times 10^{-3}$	$\frac{x}{d} = 50$	174.5	2083.5 W/m^2 K
$\ell = 1.0m$	$\hat{x} = 4.56 \times 10^{-4}$	$\frac{x}{d} = 20$	175.05	2090 W/m^2 K
$\ell = 0.5m$	$\hat{x} = 2.28 \times 10^{-4}$	$\frac{x}{d} = 10$	177.6	2120.5 W/m^2 K
$\ell = 0.25m$	$\hat{x} = 1.14 \times 10^{-4}$	$\frac{x}{d} = 5$	181.9	2171.88 W/m^2 K
$\ell = 0.1m$	$\hat{x} = 4.55 \times 10^{-5}$	$\frac{x}{d} = 2$	191.2	2283.0 W/m^2 K

We use the computer program in Problem 7.23 to obtain the Nusselt number distributions and find that $Nu_\infty \simeq 174.0$: for $x/d \geq 20$.

$$\dot{m}c_p (T_{exit} - T_{inlet}) = \dot{q}_{total} = h(\pi dL)(T_w - T_{bulk})$$

$$\therefore T_{exit} = T_{inlet} + \frac{h(\pi d\ell)}{\dot{m}c_p}(T_w - T_{bulk}) = 20°C$$

$$+ \frac{\pi d}{\dot{m}c_p}(T_w - T_{bulk}) \sum_{i=1}^{N} h_i \ell_i = 20 + 3.756 \times 10^{-4} \sum_{i=1}^{N} h_i \ell_i$$

Here h_i and ℓ_i denote the heat transfer coefficient and the pipe length over the ith segment, respectively

$$T_{exit} = 20 + 3.756 \times 10^{-4}[2283(0.1) + 2171.88(0.15)$$
$$+ 2120.5(0.25) + 2090(0.5) + 2083.5(14)]$$

$$T_{exit} = 20 + 3.756 \times 10^{-4}[228.3 + 325.8 + 530.1 + 1045$$
$$+ 29169] = 20°C + 11.75°C = 31.75°C$$

7.12 The fluid properties for $T_f = 100°C$ are: $\rho = 0.94$ kg/m^3, $c_p = 1.0011$ kJ/kg K, $\nu = 2.3 \times 10^{-6}$ m^2/s, k = 0.0318 W/m K, Pr = 0.693.

Then $R_d = \frac{u_m d}{\nu} = \frac{20 \times 0.01}{2.3 \times 10^{-6}} = 8.7 \times 10^4$

To calculate the Nusselt number we can use Fig. 7.2 for Pr = 0.72 or the Karman-Boelter-Martinelli analogy, which is

$$Nu = R_d Pr \sqrt{f/8}/\{0.833[5Pr + 5\ln(5Pr + 1) + 2.5\ln(R_d \sqrt{f/8}/6.00)]\}$$

with $f = 0.3164/R_d^{0.25}$. Both methods give Nu = 155.0. The heat flux is then

$$\dot{q}_w = \frac{k(T_w - T_m)}{d} Nu = 155 \times 0.0318 \times 7/0.01 = 3450\ W/m^2$$

7.13 Modify the computer program of problem 7.6 for the case of rough pipe to include the two-layer eddy-viscosity model defined separately in the inner and outer layers of the shear layer, and as given by Eqs. (6.83) and (7.12), respectively, with $\alpha = 0.0168$. The switch from $(\varepsilon_m)_i$ to $(\varepsilon_m)_o$ is achieved by the continuity in ε_m, namely $\varepsilon_m = (\varepsilon_m)_i$ if $(\varepsilon_m)_i \leq (\varepsilon_m)_o$ and $\varepsilon_m = (\varepsilon_m)_o$ when $(\varepsilon_m)_o > (\varepsilon_m)_i$. As in problem 7.6, the initial velocity profile is obtained by making use of the modified mixing length model. However, this time the friction factor, f, was calculated from Eq. (7.28) in order to account for the effect of the surface roughness parameter (r_o/k_s). The velocity distribution u^+ is obtained by integrating Eq. (7.13), and the velocity profile is computed until the u(y) distribution satisfies the continuity equation given by Eq. (7.18).

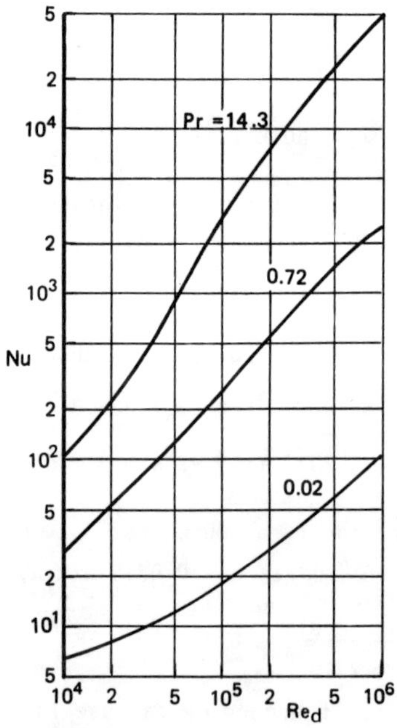

For the case of rough surface, the mixing-length expression is defined according to Eq. (6.112) with Δy expressed as a function of an equivalent sand grain-roughness parameter, k_s^+ given by Eq. (6.113).

The values of the Nusselt number obtained with the modified computer program contained in the diskette for $R_d = 10^5$ are:

Pr	$(Nu)_{Fig.\ 7.8*}$	$(Nu)_{computed}$
14.3	2882.55	2981
0.72	254.38	250
0.02	18.02	13.09

*Actually from Problem 7.22.

7.14 The Reynolds analogy factors for Pr = 0.02, 0.72, 14.3 and $R_d = 10^5$, $r_o/k_s = 100$ for both smooth and rough pipes can be obtained from Eq. (7.16) or (7.28) for f and from Fig. 7.8 for Nu are summarized below.

a. Smooth Pipe

Pr	f	Nu	St	St/(f/8)
0.02	0.0180	15.3	0.00765	3.4
0.72	0.0180	170.0	0.00236	1.05
14.3	0.0180	750.0	0.00052	0.233

b. Rough Pipe

Pr	f	Nu	St	St/(f/8)
0.02	0.0304	19.3	0.00965	2.54
0.72	0.0304	25.5	0.00354	0.932
14.3	0.0304	2800.0	0.00196	0.515

We can see that as Pr increases, the Reynolds analogy factor decreases and the Reynolds analogy (i.e. St/(f/8) = 1.0) is acceptable only for Pr close to unity.

7.15 $T_f = 0.5\ (T_e + T_w) = 0.5\ (40 + 60) = 50°C$

The fluid properties of water at $T_f = 50°C$ are: $\rho = 990\ kg/m^3$, $c_p = 4.18\ kJ/kg\ K$, $\nu = 0.568 \times 10^{-6} m^2/s$, $k = 0.640\ W/m\ K$, $\kappa = 0.153 \times 10^{-6} m^2/s$, Pr = 3.71,

Mass flow rate, $\dot{m} = \rho u_m A = 990\ u_m * \pi * (0.05^2/4) = 3.$

$$\therefore u_m = 1.543\ m/s \quad \text{and} \quad R_d = \frac{u_m d}{\nu} = \frac{0.05 \times 1.543}{0.568 \times 10^{-6}} = 1.36 \times 10^5$$

$$f = \frac{1}{[2\log_{10}(r_o/k_s) + 1.74]^2} = \frac{1}{[2\log_{10}(100) + 1.74]^2} = 0.03035$$

To compute St, we use Reynolds analogy, St/(f/8) = 1.0

$\therefore$ St = f/8 = 0.00379 so that

$$Nu = Pr\ R_d\ St = 3.71 \times 1.36 \times 10^5 \times 0.00379 = 1914$$

and $\dot{q}_w = Nu\ (T_w - T_m)\ \frac{k}{d}\ \pi dL = (1914)(10)(0.640)\pi(20) = 7.7 \times 10^5$ W

7.16 The fluid properties of air at $T = 10°C$ are $\rho = 1.257$ kg/m^3, c_p = 1.0055 kJ/kgK, $\nu = 1.36 \times 10^{-5}$ m^2/s, k = 0.0249 W/mK, Pr = 0.72. The friction factor is calculated from the formula

$$f = \frac{1}{[2\log_{10}(r_0/k_s) + 1.74]^2} = 0.0234 \quad \text{and } R_d \text{ is}$$

$$R_d = \frac{ud}{\nu} = \frac{\dot{m}\,d}{\rho A \nu} = \frac{0.5 \times 0.02}{1.257 \times \pi \times (0.01)^2 \times 1.36 \times 10^{-5}} = 1.87 \times 10^6$$

Assume that the Reynolds analogy is valid, i.e., St/(f/8) = 1.0, so that $Nu = PrR_dSt = 0.72 \times 1.87 \times 10^6 \times (0.0234/8) = 3.94 \times 10^3$, and $\dot{q}_w = Nu(T_w - T_m)\ k/d = 3.94 \times 10^3 \times 5 \times 0.0249/0.02 = 2.45 \times 10^4$ W/m^2. The heat transfer per unit length of the pipe is then

$$\dot{Q}_w = \dot{q}_w \cdot 2\pi r = 2.45 \times 10^4 \times 2\pi \times 0.01 = 1.54 \text{ W/m.}$$

7.17

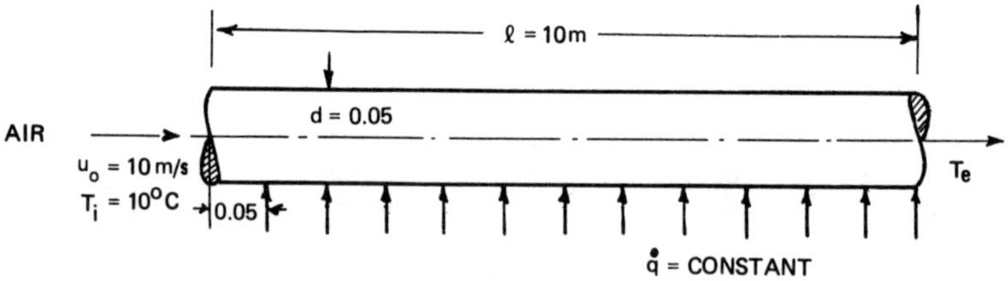

Assume fully developed velocity field and evaluate the properties of air at 10°C. $\nu = 12.58 \times 10^{-6}$ m^2/s; $k \simeq 0.02624$ W/mK then $R_{r_0} = \frac{u_0 r_0}{\nu}$

$$= \frac{(10\text{m/s})(2.5 \times 10^{-2})\text{m}}{12.58 \times 10^{-6}\ \text{m}^2/\text{s}} = 20 \times 10^3.$$

With constant heat flux boundary condition, Eqs. (P7.2) and (P7.3), with $Nu_\infty = 88.47$ and $R_d = 40 \times 10^3$:

$$\frac{Nu}{Nu_\infty} = 1.0 + 0.8(1.0 + 70{,}000R_d^{-1.5})(\tfrac{x}{d})^{-1}, \quad Nu_\infty = (0.021)(R_d)^{0.8}(Pr)^{0.4}$$

$\frac{x}{d}$	$\frac{Nu}{Nu_\infty}$	Nu	$\bar{h} = \frac{Nuk}{d}$	$\dot{q}_w = \bar{h}(T_w - \bar{T})$
$\frac{8}{5} = 1.6$	1.50	132.7	69.64 W/m^2°C	348.2 W/m^2
$\frac{20}{5} = 4.0$	1.20	106.2	55.7 W/m^2°C	278.5 W/m^2
$\frac{40}{5} = 8.0$	1.10	97.3	51.1 W/m^2°C	255.4 W/m^2

7.18 Evaluate the properties of air at $\bar{T}$ = 200°C (473K). $Pr = 0.680$; $\nu = 53.38 \times 10^{-6}\ m^2/s$; $k = 0.0387$ W/m K; $\rho = 0.744\ kg/m^3$; $c_p = 1.024\ \frac{kJ}{kg\ K}$

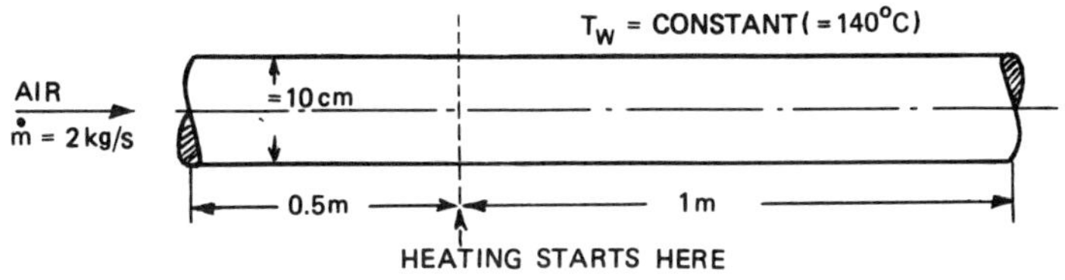

Since $\dot{m} = \rho\bar{u}\,\frac{\pi d^2}{4}$, $\quad \bar{u} = \frac{4\dot{m}}{\rho\pi d^2} = \frac{4 \times 2}{0.744 \times (\pi)(0.1)^2} = 3.42 \times 10^2\ ms^{-1}$

$$R_{r_0} = \frac{\bar{u}r_0}{\nu} = \frac{(342)m/s\ (0.05)}{53.38 \times 10^{-6}} \simeq 512{,}202; \text{ Take } R_{r_0} \sim 500{,}000 \text{ or } R_d = 10^6$$

At this Reynolds number, from Fig. (7.2), $Nu_\infty \to 1150$. Also use Fig. 7.9b to obtain the variation of $\frac{Nu}{Nu_\infty}$ with $\frac{x}{d}$; after the <u>heating starts</u>. From Fig. 7.9b; at $R_d = 10^6$,

$$\frac{Nu}{Nu_\infty}\left(\frac{x}{d} = \frac{0.3}{0.1}\right) \to \frac{Nu}{Nu_\infty}\left(\frac{x}{d} = 3\right) = 1.2, \quad Nu = 1380; \quad h = 534.1\ W/m^2\ K$$

$$\frac{Nu}{Nu_\infty}\left(\frac{x}{d} = 4\right) \to \frac{Nu}{Nu_\infty}\left(\frac{x}{d} = 4\right) = 1.15, \quad Nu = 1322.5; \quad h = 511.8\ W/m^2\ K$$

$$\frac{Nu}{Nu_\infty}\left(\frac{x}{d} = 6\right) \to \frac{Nu}{Nu_\infty}\left(\frac{x}{d} = 6\right) = 1.10, \quad Nu = 1265; \quad h = 490\ W/m^2\ K$$

The local heat transfer rates are then:

$\frac{x}{d}$	$h[W/m^2\ K]$	$\dot{q}_w = h(T_w - \bar{T})\ [W/m^2]$	ΔT (temperature drop)
3	534	-32040	-1.47°C
4	512	-30720	-0.47°C
6	490	-29400	-0.90°C

The decrease in air temperature is obtained from

$$\dot{m}c_p\,[T_i - T(x)] = h\pi d(x)(T_w - \bar{T}), \quad T_i - T(x) = \frac{h\pi dx(T_w - \bar{T})}{\dot{m}c_p}$$

or $\quad \Delta T_1 = T(x = 0.3) - T_i = \frac{(-32040)(\pi)(0.1)(0.3)}{(2)(1024)} = -1.47°C.$

Similarly $\Delta T_2 = -0.47°C$, $\Delta T_3 = -0.90°C$.

$\Delta T_{total} = -2.84°C$, total temperature drop $\simeq$ 3°C

7.19 Use the Leveque solution given in Problem 4.20 where

$$Nu_x = \frac{x}{0.893}\left(\frac{\lambda Pr}{9\nu x}\right)^{1/3} \quad \text{with} \quad \lambda = \left(\frac{\partial u}{\partial y}\right)_w.$$

Since near wall, $u^+ = y^+$ and $du^+/dy^+ = 1$

$$\left(\frac{\partial u}{\partial y}\right)_w = \frac{u_\tau^2}{\nu} = \frac{1}{\nu}\left(\frac{\tau_w}{\rho}\right)$$

From the definition of friction factor f and Eq. (7.17), it follows that

$$\left(\frac{\partial u}{\partial y}\right)_w = \frac{1}{\nu}\frac{f}{8}u_m^2 = \frac{1}{\nu}\frac{0.3164}{8\,R_d^{1/4}}u_m^2$$

Substituting the above expression into the Nu_x-expression, with $\bar{x} = \frac{x}{d}$,

$$Nu_x = \frac{\bar{x}d}{0.893}\left[\frac{1}{\nu^2}\frac{Pr}{9\bar{x}d}\frac{0.3164}{8\,R_d^{1/4}}u_m^2\right]^{1/3}$$

and rearranging

$$Nu_x = \frac{1}{0.893}\left[0.0044\frac{R_d^2\,Pr\bar{x}^1}{R_d^{1/4}}\right]^{1/3}$$

Substituting the numerical values

$$Nu_x = \frac{1}{0.893}\left[0.0044\frac{(10^4)^2(100)\bar{x}^1}{(10^4)^{1/4}}\right]^{1/3} \simeq 187\ \bar{x}^{1/3}$$

Since, with ℓ_t denoting the entrance length,

$$\overline{Nu} = \frac{1}{\ell_t}\int_0^{\ell_t} Nu(x)dx = \frac{d}{\ell_t}(187)\left.\frac{(x^+)^{2/3}}{2/3}\right|_0^{\ell_t/d}$$

$$= 281\left(\frac{\ell_t}{d}\right)^{1/3}$$

From Fig. 7.5, $\overline{Nu} \simeq 200$

$\therefore\ \ell_t/d \simeq 22.$

7.20 From the computer program in the diskette, (a) the variation of the local Nusselt number, and (b) the temperature profiles at x/d = 2,4,6 and 8 are given below for $R_d = 10^5$, Pr = 14.3.

(a)

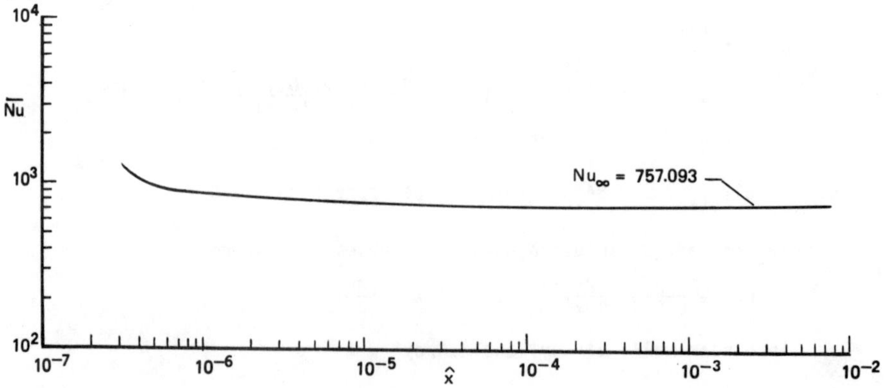

(b)

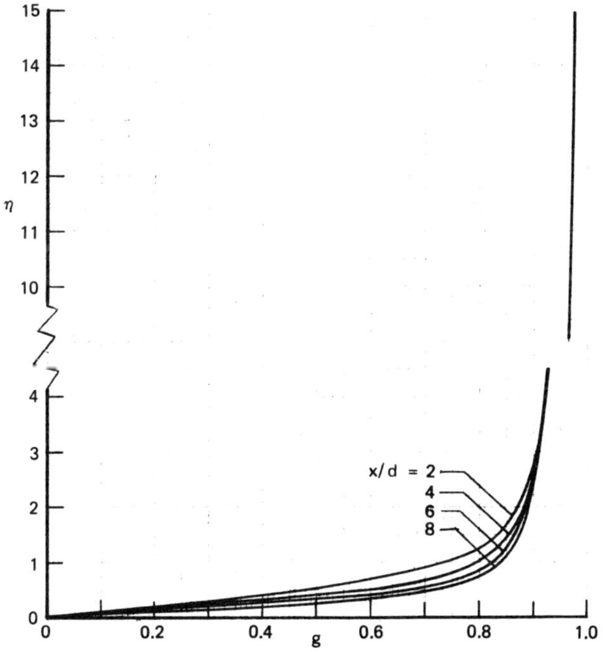

7.21 From the computer program in the diskette, comparison of computed results with those from problem 7.13 are shown below.

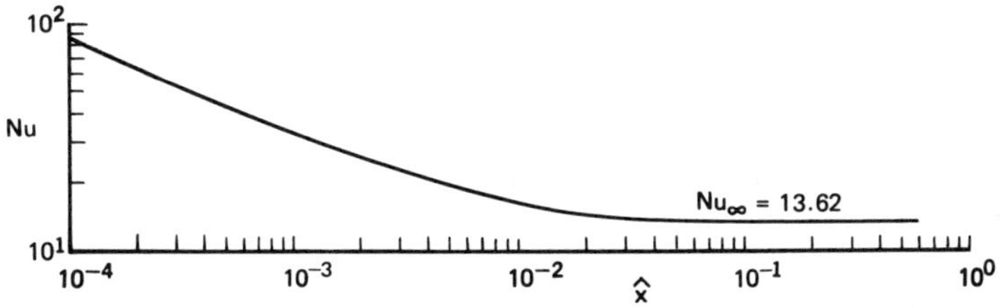

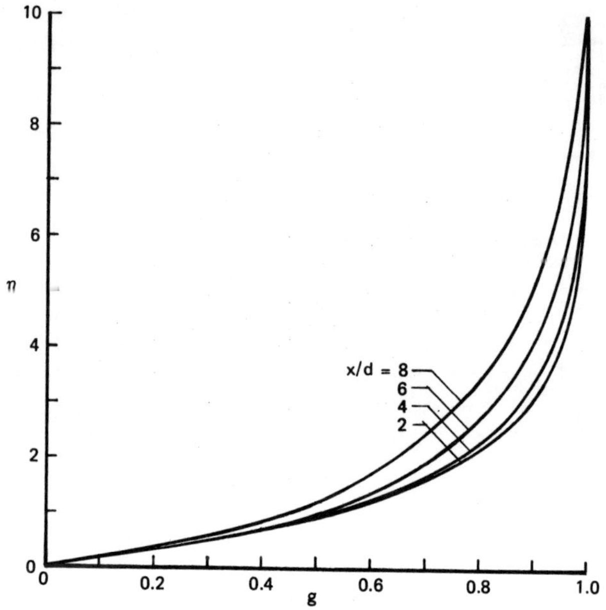

7.22 To include the effect of surface roughness for turbulent pipe flow with constant wall temperature, the computer program of Section 13.3 is modified. A new subroutine called "EDDY" is introduced which calculates the two-layer eddy viscosity as discussed in problem (7.13) (see the diskette). For a surface roughness of $r_o/k_s = 100$; $R_d = 10^5$, Pr = 0.02, 0.72, and 14.3, the following Nusselt numbers are obtained:

Pr = 14.3	Pr = 0.72	Pr = 0.02
Nu = 2981	Nu = 250	Nu = 13.09

The results for a fully-developed thermal boundary layer (Problem 7.13) are also given for comparison:

Nu = 2882.55	Nu = 254.38	Nu = 18.02

7.23 The computer program in Section 13.3 was modified for constant heat flux boundary condition, as in problem (5.18) and is presented in the diskette. The calculations indicate that $Nu_\infty \to 171$ as $\hat{x} \to \infty$. Figures below show the variation of Nu as a function of $\hat{x}$ $(= \frac{x/r_o}{R_d Pr})$ and $\frac{Nu}{Nu_\infty}$ as a function of $\frac{x}{d}$, respectively, together with the predictions of Eqs. (P7.2) and (P7.3).

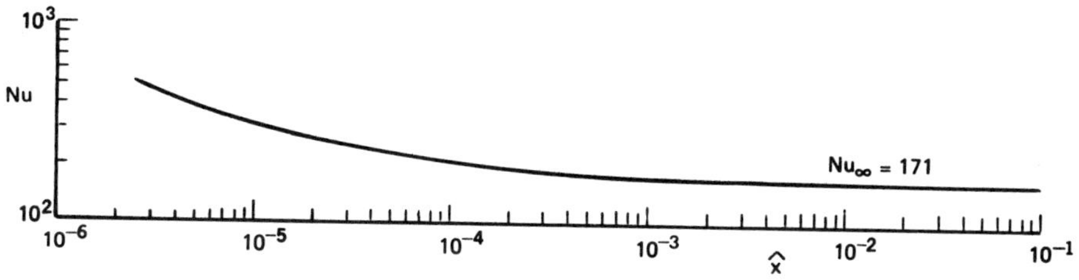

10^3
Nu
10^2
10^-6
10^-5
10^-4
10^-3
x̂
10^-2
10^-1
Nu∞ = 171

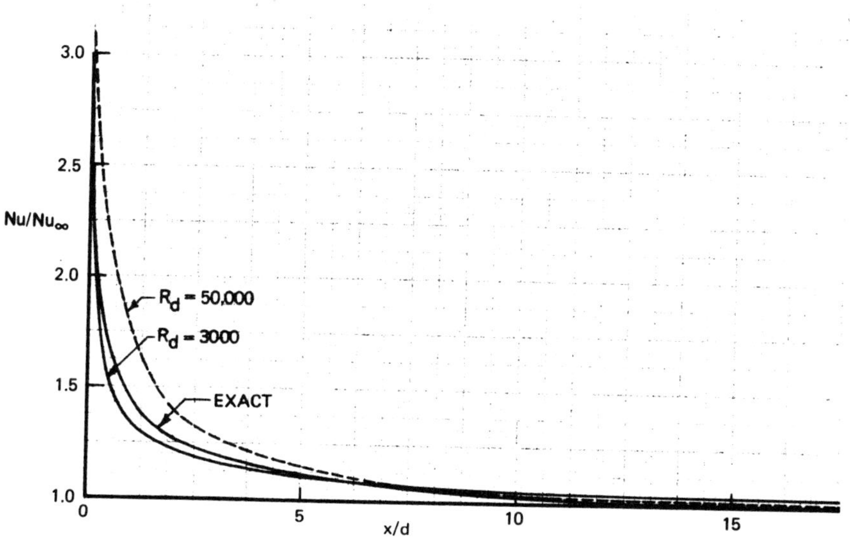

3.0
2.5
Nu/Nu∞
2.0
1.5
1.0
0
5
10
15
x/d
Rd = 50,000
Rd = 3000
EXACT

Chapter 8

Free Shear Flows

8.1 (a) With $u = \frac{1}{r}\frac{\partial\psi}{\partial r}$, $v = -\frac{1}{r}\frac{\partial\psi}{\partial x}$, the momentum equation becomes

$$\frac{1}{r}\frac{\partial\psi}{\partial r}\frac{\partial}{\partial x}\left(\frac{1}{r}\frac{\partial\psi}{\partial r}\right) - \frac{1}{r}\frac{\partial\psi}{\partial x}\frac{\partial}{\partial r}\left(\frac{1}{r}\frac{\partial\psi}{\partial r}\right) = \frac{\nu}{r}\frac{\partial}{\partial r}\left\{r\frac{\partial}{\partial r}\left(\frac{1}{r}\frac{\partial\psi}{\partial r}\right)\right\}$$

With $x = A^{\alpha_1}\bar{x}$, $r = A^{\alpha_2}\bar{r}$, $\psi = A^{\alpha_3}\bar{\psi}$, we have

$$A^{2\alpha_3-4\alpha_2-\alpha_1}\left\{\frac{1}{\bar{r}}\frac{\partial\bar{\psi}}{\partial\bar{r}}\frac{\partial}{\partial\bar{x}}\left(\frac{1}{\bar{r}}\frac{\partial\bar{\psi}}{\partial\bar{r}}\right)\right\} - A^{2\alpha_3-4\alpha_2-\alpha_1}\left\{\frac{1}{\bar{r}}\frac{\partial\bar{\psi}}{\partial\bar{x}}\frac{\partial}{\partial\bar{r}}\left(\frac{1}{\bar{r}}\frac{\partial\bar{\psi}}{\partial\bar{r}}\right)\right\}$$

$$= A^{\alpha_3-4\alpha_2}\frac{\nu}{\bar{r}}\frac{\partial}{\partial\bar{r}}\left\{\bar{r}\frac{\partial}{\partial\bar{r}}\left(\frac{1}{\bar{r}}\frac{\partial\bar{\psi}}{\partial\bar{r}}\right)\right\}$$

Invariance under the transformation requires that

$$2\alpha_3 - 4\alpha_2 - \alpha_1 = 2\alpha_3 - 4\alpha_2 - \alpha_1 = \alpha_3 - 4\alpha_2$$

from which, $\alpha_3/\alpha_1 = 1$, $\alpha_2/\alpha_1 = 1$ implying that the similarity variable η and the stream function ψ can be written as

$$\eta = \frac{r}{x}, \quad \psi = xf(\eta) \quad \text{or} \quad \psi = \nu xf(\eta) \text{ (dimensionally correct)}$$

(b) $$u = \frac{1}{r}\frac{\partial\psi}{\partial r} = \frac{1}{r}\nu xf'\frac{\partial\eta}{\partial r} = \frac{\nu f'}{r} = \frac{f'\nu}{\eta x} \quad \therefore \quad \frac{u}{u_c} = \frac{1}{u_c x/\nu}\frac{f'}{\eta}$$

(c) For $\eta = \left(\frac{u_c x}{\nu}\right)^{1/2}\frac{r}{x}$, $\psi = \nu xf(\eta)$,

$$u = \frac{1}{r}\frac{\partial\psi}{\partial r} = \frac{1}{r}\frac{\partial\psi}{\partial\eta}\frac{\partial\eta}{\partial r} = \nu xf'\frac{\eta}{r^2} = \left(\frac{u_c x}{\nu}\right)\frac{\eta}{\eta^2 x^2}\nu xf' = u_c\frac{f'}{\eta}$$

By the chain rule,

$$\left(\frac{\partial u}{\partial x}\right)_r = \left(\frac{\partial u}{\partial x}\right)_\eta + \frac{\partial u}{\partial\eta}\frac{\partial\eta}{\partial x} = \frac{du_c}{dx}\frac{f'}{\eta} + u_c\left(\frac{f'}{\eta}\right)'\frac{\partial\eta}{\partial x}$$

$$-v = \frac{1}{r}\frac{\partial\psi}{\partial x} = -\left(\frac{u_c x}{\nu}\right)^{1/2}\frac{\nu}{x}\left[\frac{f}{\eta} + x\frac{f'}{\eta}\frac{\partial\eta}{\partial x}\right]$$

$$u \frac{\partial u}{\partial x} = u_c \frac{du_c}{dx} \left(\frac{f'}{\eta}\right)^2 + u_c^2 \frac{f'}{\eta} \left(\frac{f'}{\eta}\right)' \frac{\partial \eta}{\partial x}$$

$$-v \frac{\partial u}{\partial r} = -v \frac{\partial u}{\partial \eta} \frac{\partial \eta}{\partial r} = \left(\frac{u_c x}{\nu}\right)^{1/2} \frac{\nu}{x} \left[\frac{f}{\eta} + x \frac{f'}{\eta} \frac{\partial \eta}{\partial x}\right] \frac{1}{x} \left(\frac{u_c x}{\nu}\right)^{1/2} \left(\frac{f'}{\eta}\right)' u_c$$

$$= \frac{u_c^2}{x} \left[\frac{f}{\eta} \left(\frac{f'}{\eta}\right)' + x \frac{f'}{\eta} \left(\frac{f'}{\eta}\right)' \frac{\partial \eta}{\partial x}\right]$$

$$\nu \frac{1}{r} \frac{\partial}{\partial r} \left(r \frac{\partial u}{\partial r}\right) = \nu \frac{1}{r} \frac{\partial}{\partial \eta} \left(r \frac{\partial u}{\partial \eta} \frac{\partial \eta}{\partial r}\right) \frac{\partial \eta}{\partial r} = \nu \frac{\eta}{r^2} \left[\eta \left(\frac{f'}{\eta}\right)'\right]' u_c$$

$$= \frac{u_c^2}{x} \frac{1}{\eta} \left[\eta \left(\frac{f'}{\eta}\right)'\right]'$$

$$u \left(\frac{\partial T}{\partial x}\right)_r = u \left(\frac{\partial T}{\partial x}\right)_\eta + u \frac{\partial T}{\partial \eta} \frac{\partial \eta}{\partial x} = u_c \frac{f'}{\eta} \left[G \frac{\partial T_c}{\partial x} + (T_c - T_e) G' \frac{\partial \eta}{\partial x}\right]$$

$$v \frac{\partial T}{\partial r} = v \frac{\partial T}{\partial \eta} \frac{\partial \eta}{\partial x} = - \left(\frac{u_c x}{\nu}\right)^{1/2} \frac{\nu}{x} \left[\frac{f}{\eta} + x \frac{f'}{\eta} \frac{\partial \eta}{\partial x}\right] (T_c - T_e) G' \left(\frac{u_c x}{\nu}\right)^{1/2} \frac{1}{x}$$

$$= - \frac{u_c}{x} (T_c - T_e) \left[G' \frac{f}{\eta} + x \frac{f'}{\eta} \frac{\partial \eta}{\partial x} G'\right]$$

$$\frac{\nu}{Pr} \frac{1}{r} \frac{\partial}{\partial r} \left(r \frac{\partial T}{\partial r}\right) = \frac{\nu}{Pr} \left(\frac{\eta G'}{r}\right)' (T_c - T_e) = \frac{u_c}{x} (T_c - T_e) \frac{1}{Pr} (\eta G')' / \eta$$

Substituting the above relations into (P8.2) and (P8.3)

$$u_c \frac{du_c}{dx} \left(\frac{f'}{\eta}\right)^2 - \frac{u_c^2}{x} \frac{f}{\eta} \left(\frac{f'}{\eta}\right)' = \frac{u_c^2}{x} \frac{1}{\eta} \left[\eta \left(\frac{f'}{\eta}\right)'\right]' \qquad (1)$$

$$u_c \frac{dT_c}{dx} G \frac{f'}{\eta} - \frac{u_c}{x} (T_c - T_e) G' \frac{f}{\eta} = \frac{u_c}{x} (T_c - T_e) \frac{1}{Pr} \left(\frac{\eta G'}{\eta}\right)' \qquad (2)$$

with $u_c x$ = constant and $(T_c - T_e)\, x$ = constant we get

$$u_c \frac{du_c}{dx} = - \frac{u_c^2}{x}, \quad u_c \frac{dT_c}{dx} = - u_c (T_c - T_e)/x.$$

Equations (1) and (2) and their b. c. can then be written as

$$\left[\eta \left(\frac{f'}{\eta}\right)'\right]' + f \left(\frac{f'}{\eta}\right)' + \frac{(f')^2}{\eta} = 0 \qquad (P8.12)$$

$$\left[\frac{\eta}{Pr} G' + fG\right]' = 0 \qquad (P8.13)$$

$$\eta = 0, \quad f = G' = 0, \quad f'' = 0; \quad \eta = \eta_e, \quad f' = G = 0$$

(d) Since $\left(\frac{ff'}{\eta}\right)' = f' \frac{f'}{\eta} + f \left(\frac{f'}{\eta}\right)'$, Eq. (P8.12) becomes

$$\left[\eta \left(\frac{f'}{\eta}\right)'\right]' + \left(\frac{ff'}{\eta}\right)' = 0$$

Integrating wrt η twice,

$$\eta \left(\frac{f'}{\eta}\right)' + \frac{ff'}{\eta} = \text{const}, \quad f'' + \frac{ff'}{\eta} - \frac{f'}{\eta} = c_1 \qquad (1)$$

but $c_1 = 0$ from $\lim_{\eta \to \infty} \frac{f'(\eta)}{\eta} = 0$ and $\lim_{\eta \to \infty} f''(\eta) = 0$

The second Eq. in (1) can be rewritten as, $(\frac{f'}{\eta})' + \frac{(f^2)'}{2\eta^2} = 0$ and its solution subject to f(0) = 0 is,

$$f(\eta) = \frac{(1/2)\eta^2}{1 + (1/8)\eta^2}$$

(e) Integrating (P8.13) wrt η and with f = G' = 0 at η = 0,

$$\frac{\eta}{Pr} G' + fG = 0 \quad \text{or} \quad \frac{G'}{G} = - \frac{Pr\ f}{\eta} = - \frac{Pr\ (1/2)\eta}{1 + (1/8)\eta^2}$$

With $z = 1 + \frac{1}{8}\eta^2$, $dz = \frac{1}{4}\eta d\eta$, $\frac{dG}{G} = - \frac{2Pr}{z} dz$

$$\therefore \quad G = Az^{-2Pr} = A \frac{1}{(1 + (1/8)\eta^2)^{2Pr}}$$

Since η = 0, G = 1 $\therefore$ A = 1 $\therefore$ $G = \frac{1}{(1 + 1/8\ \eta^2)^{2Pr}}$

8.2 u_c, $\frac{\delta}{h}$ and $\dot{m}/\rho$ for a 2-D laminar jet follow from (8.29a-c) in the book.

(a) Parabolic velocity profile: $u = u_{max}[1 - (y/h)^2]$;

$$\frac{J}{\rho} = 2 \int_0^X u^2 dy = 2u_{max}^2 \int_0^h [1 - (\frac{y}{h})^2]^2 dy = \frac{16}{15} u_{max}^2 h$$

$$u_c = [\frac{3}{32} (\frac{16}{15} \times 0.5^2 \times 0.05)^2 \frac{1}{1.5 \times 10^{-5} \times 0.05}]^{1/3} (\frac{h}{x})^{1/3}$$

$$= 2.81 (\frac{h}{x})^{1/3} \text{ m/s}$$

$$\frac{\delta}{h} = [\frac{12\sqrt{2}\ \nu^2 x^2}{16/15\ u_{max}^2 h^4}]^{1/3} = [\frac{12\sqrt{2}\ \nu^2}{16/15 \times 0.5^2 \times 0.05^2}]^{1/3} (\frac{x}{h})^{2/3}$$

$$= 1.79 \times 10^{-2} (\frac{x}{h})^{2/3}$$

$$\frac{\dot{m}}{\rho} = 3.302 (\frac{J}{\rho})^{1/3} (\nu h)^{1/3} (\frac{x}{h})^{1/3} \text{ m}^3/\text{s}$$

$$= 3.302 (\frac{16}{15} \times 0.5^2 \times 0.05)^{1/3} (1.5 \times 10^{-5} \times 0.05)^{1/3} (\frac{x}{h})^{1/3}$$

$$= 7.114 \times 10^{-3} (\frac{x}{h})^{1/3}$$

(b) Uniform flow: $u = u_{max}$ = const: $\frac{J}{\rho} = 2 \int_0^\infty u^2 dy = 2\ u_{max}^2 h$

$$u_c = [\frac{3}{32} \times (2 \times 0.5^2 \times 0.05)^2 \frac{1}{1.5 \times 10^{-5} \times 0.05}]^{1/3} (\frac{h}{x})^{1/3}$$

$$= 4.275 (\frac{h}{x})^{1/3} \text{ m/s}$$

$$\frac{\delta}{h} = [\frac{12\sqrt{2} \times (1.5 \times 10^{-5})^2}{2 \times 0.5^2 \times 0.05^2}]^{1/3} (\frac{x}{h})^{2/3} = 1.451 \times 10^{-2} (\frac{x}{h})^{2/3}$$

$$\frac{\dot{m}}{\rho} = 3.302\ [2 \times 0.5^2 \times 0.05 \times 1.5 \times 10^{-5} \times 0.05]^{1/3} \left(\frac{x}{h}\right)^{1/3}$$

$$= 8.77 \times 10^{-3} \left(\frac{x}{h}\right)^{1/3}\ m^3/s$$

8.3 For a similarity solution of Eq. (8.58), it is necessary that

$$\frac{Pr_t}{\varepsilon_m} \frac{u_c \delta^2}{T_c - T_e} \frac{d}{dx} (T_c - T_e) = \text{const} = -cPr_t \qquad (1)$$

$$\frac{u_c \delta}{\varepsilon_m} = \frac{2c_1}{d\delta/dx} \qquad (2)$$

Substituting (2) into (1) we get

$$\frac{2c_1}{d\delta/dx} \frac{\delta}{T_c - T_e} \frac{d}{dx} (T_c - T_e) = -c \qquad (3)$$

or $\dfrac{1}{T_c - T_e} \dfrac{d}{dx} (T_c - T_e) = \dfrac{A}{\delta} \dfrac{d\delta}{dx}$

Here $A = -c/2c_1$ = const = $-1/2$ for $c = c_1$. Integrating (3) and taking $A = -1/2$

$$\ell n\,(T_c - T_e) = \ell n \delta^{-1/2} + B \quad \text{or} \quad T_c - T_e = B_1 \delta^{-1/2}$$

where B_1 is determined from the initial conditions.

8.4 With $u_d = u_1 - u \ll u_1$, $u \dfrac{\partial u}{\partial x} = -(u_1 - u_d) \dfrac{\partial u_d}{\partial x} \approx -u_1 \dfrac{\partial u_d}{\partial x}$

$$v \frac{\partial u}{\partial y} = -v \frac{\partial u_d}{\partial y} \approx -u_d \frac{\partial u_d}{\partial x} \ll -u_1 \frac{\partial u_d}{\partial x}, \quad \nu \frac{\partial^2 u}{\partial y^2} = -\nu \frac{\partial^2 u_d}{\partial y^2}$$

With the above approximations, the momentum equation and its b.c. becomes

$$u_1 \frac{\partial u_d}{\partial x} = \nu \frac{\partial^2 u_d}{\partial y^2}; \quad y \to \infty:\ u_d = 0; \quad y \to -\infty:\ u_d = u_1 - u_2 \qquad (1)$$

To obtain a similarity solution, we let $u_d = f(\eta)$, $\eta = y/\delta(x)$.

$$\frac{\partial u_d}{\partial x} = \frac{\partial f}{\partial \eta} \frac{\partial \eta}{\partial x} = -\eta f' \frac{1}{\delta(x)} \frac{d\delta(x)}{dx}, \quad \frac{\partial^2 u_d}{\partial y^2} = f'' \frac{1}{\delta^2(x)} \qquad (2)$$

The first eq. in (1) becomes $-\dfrac{u_1 \delta}{\nu} \dfrac{d\delta}{dx} \eta f' = f''$ (3)

but similarity requires that $\dfrac{u_1 \delta}{\nu} \dfrac{d\delta}{dx} = c$ (4)

Take $c = 1/2$ and integrate to get $\delta = \left(\dfrac{\nu x}{u_1}\right)^{1/2}$. Substituting (4) into (3) and integrating twice

$$u_d = c_1 \int_{-\infty}^{\eta} e^{-\eta^2/4} d\eta + c_2 \qquad (5)$$

where $c_2 = u_1 - u_2$, $c_1 = -(u_1 - u_2)/\sqrt{4\pi}$. To satisfy b.c. in (1), write

$$u_1 - u = (u_1 - u_2) - \frac{(u_1 - u_2)}{2\sqrt{\pi}} \int_{-\infty}^{\eta} e^{-\eta^2/4} d\eta$$

$$- \frac{u_1 - u_2}{2} - \frac{u_1 - u_2}{2\sqrt{\pi}} \int_0^{\eta} e^{-\eta^2/4} d\eta = \frac{u_1 - u_2}{2} [1 - \frac{2}{\sqrt{\pi}} \int_0^{\eta/2} e^{-\eta^2} d\eta]$$

or $$\frac{u}{u_1} = 1 - \frac{1}{2} (1 - \frac{u_2}{u_1})[1 - \mathrm{erf}\,(\frac{\eta}{2})]$$

Then $$\tau_o = \mu (\frac{\partial u}{\partial y})_{y=0} = \mu (\frac{\partial u}{\partial \eta})(\frac{\partial \eta}{\partial y})_{\eta=0}$$

$$= \mu u_1 [- \frac{1}{2} (1 - \frac{u_2}{u_1})(- \frac{2}{\sqrt{\pi}}) e^{-\eta^2/4}]_{\eta=0} (\frac{u_1}{\nu x})^{1/2}$$

or $$c_f = \frac{\tau_o}{1/2 \, \rho u_1^2} = (\frac{\nu}{u_1 x})^{1/2} \frac{2}{\sqrt{\pi}} (1 - \frac{u_2}{u_1})$$

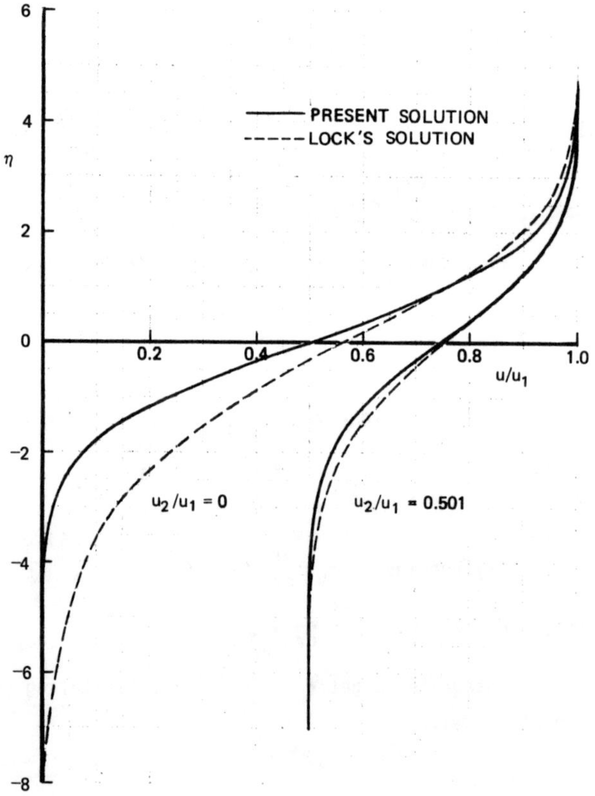

The present approximate solutions and Lock's exact solutions for $u_2/u_1 = 0$ and 0.501 are shown in the figure. For $u_2/u_1 = 0.5$, the agreement is good, particularly in the upper branch, despite the assumption that $u_d << u_1$. For $u_2/u_1 = 0$, the agreement in the upper branch ($\eta > 0$) is still satisfactory whereas in the lower branch, the agreement is poor. In the latter case, the assumption $u_d = u_1 - u << u_1$ is obviously violated.

8.5 (a) $$\lim_{Pr \to 0} K = \lim_{Pr \to 0} 2\sqrt{2}\, \rho c_p N \int_0^\infty \frac{d\eta/\sqrt{2}}{\cosh^{2Pr+2}(\eta/\sqrt{2})}$$

$$= 2\sqrt{2}\, \rho c_p N \int_0^\infty \frac{d\eta/\sqrt{2}}{\cosh^2(\eta/\sqrt{2})} = 2\sqrt{2}\, \rho c_p N \quad \therefore \quad N = \frac{1}{2\sqrt{2}} \frac{K}{\rho c_p}$$

(b) For $Pr = 0.5$, $$K = 2\sqrt{2}\, \rho c_p N \int_0^\infty \frac{d\eta/\sqrt{2}}{\cosh^3(\eta/\sqrt{2})}$$

Chapter 9

Buoyant Flows

9.1 By the matrix elimination method, construct a matrix whose columns give the mass, length, time, and temperature dimensions of the variables in each row

	M	L	T	θ
h	1	0	-3	-1
ρ	1	-3	0	0
μ	1	-1	-1	0
c_p	0	2	-2	-1
k	1	1	-3	-1
L	0	1	0	0
$g\beta$	0	1	-2	-1
ΔT	0	0	0	1

First eliminate the dimension [M] by dividing the variables containing the dimension [M] by ρ except h, which is divided by k. Next, eliminate the dimension [θ] by forming the groups $g\beta\Delta T$, $k/\rho c_p$

	L	T
h/k	-1	0
μ/ρ	2	-1
$k/\rho c_p$	2	-1
L	1	0
$g\beta\Delta T$	1	-2

The dimension time [T] is then eliminated by forming the groups $\rho^2 g\Delta T/\mu^2$ and $\mu c_p/k$. Finally, by inspection, the length scale is eliminated by forming the groups, hL/k, $L^3\rho^2 g\beta\Delta T/\mu^2$.

$$\frac{hL}{k} = f\ (Pr,\ L^3\rho^2 g\beta\Delta T/\mu^2) \tag{1}$$

Since $Lg\beta\Delta T$ has the dimensions of L^2/T^2, which is equivalent to $[u]^2$, we introduce a velocity scale u_c given by $u_c = \sqrt{g\beta L\Delta T}$, so that

$$\frac{\rho^2 L L^2 g\beta\Delta T}{\mu^2} = \frac{L^2\rho^2 u_c^2}{\mu^2} = Gr_L \tag{2}$$

Nusselt number hL/k and Eq. (1) can then be expressed as

$$Nu = f\ (Gr_L,\ Pr) \tag{3}$$

9.2 From the definition $\beta = -\frac{1}{\rho}(\frac{\partial\rho}{\partial T})_p$, we can write

$$\Delta T\beta = -\frac{\Delta\rho}{\rho} \quad \text{or} \quad (T_w - T_e)\beta = -\frac{(\rho_e - \rho_w)}{\rho}$$

Substituting the above relation into $\frac{u_c}{u_e} = \frac{\sqrt{g\beta L(T_w - T_e)}}{u_e}$, we get $(\frac{u_c}{u_e})^2 = \frac{\Delta\rho}{\rho}\frac{gL}{u_e^2}$, which, by definition, is the Richardson number.

9.3 By introducing the usual definition of stream function, ψ, and neglecting the terms due to turbulence and pressure gradient, we can write the momentum and energy equations (9.12) and (9.13) as

$$\frac{\partial\psi}{\partial y}\frac{\partial^2\psi}{\partial x\partial y} - \frac{\partial\psi}{\partial x}\frac{\partial^2\psi}{\partial y^2} = \nu\frac{\partial^3\psi}{\partial y^3} + g\beta(T - T_e), \tag{1a}$$

$$\frac{\partial\psi}{\partial y}\frac{\partial T}{\partial x} - \frac{\partial\psi}{\partial x}\frac{\partial T}{\partial y} = \frac{\nu}{Pr}\frac{\partial^2 T}{\partial y^2} \tag{1b}$$

We then introduce the linear transformation defined by

$$x = A^{\alpha_1}\bar{x}, \quad y = A^{\alpha_2}\bar{y}, \quad \psi = A^{\alpha_3}\bar{\psi}, \quad T = A^{\alpha_4}\bar{T}, \quad T_e = A^{\alpha_4}\bar{T}_e$$

and write Eqs. (1) as

$$A^{2\alpha_3-\alpha_1-2\alpha_2}(\frac{\partial\bar{\psi}}{\partial\bar{y}}\frac{\partial^2\bar{\psi}}{\partial\bar{x}\partial\bar{y}} - \frac{\partial\bar{\psi}}{\partial\bar{x}}\frac{\partial^2\bar{\psi}}{\partial\bar{y}^2}) = A^{\alpha_3-3\alpha_2}\nu\frac{\partial^3\bar{\psi}}{\partial\bar{y}^3} + A^{\alpha_4}g\beta(\bar{T} - \bar{T}_e)$$

$$A^{\alpha_3+\alpha_4-\alpha_1-\alpha_2}(\frac{\partial\bar{\psi}}{\partial\bar{y}}\frac{\partial\bar{T}}{\partial\bar{x}} - \frac{\partial\bar{\psi}}{\partial\bar{x}}\frac{\partial\bar{T}}{\partial\bar{y}}) = A^{\alpha_4-2\alpha_2}\frac{\nu}{Pr}\frac{\partial^2\bar{T}}{\partial\bar{y}^2}$$

The invariance under the transformation requires that

$$2\alpha_3 - \alpha_1 - 2\alpha_2 = \alpha_3 - 3\alpha_2 = \alpha_4, \quad \alpha_3 + \alpha_4 - \alpha_1 - \alpha_2 = \alpha_4 - 2\alpha_2$$

which result in $\frac{\alpha_3}{\alpha_1} = 1 - \alpha$, $\frac{\alpha_4}{\alpha_1} = 1 - 4\alpha$ where $\alpha = \frac{\alpha_2}{\alpha_1}$.

so that $\frac{y}{x^\alpha} = \frac{\bar{y}}{\bar{x}^\alpha}$; $\frac{\psi}{x^{1-\alpha}} = \frac{\bar{\psi}}{\bar{x}^{1-\alpha}}$; $\frac{T}{x^{1-4\alpha}} = \frac{\bar{T}}{\bar{x}^{1-4\alpha}}$; $\frac{T_e}{x^{1-4\alpha}} = \frac{\bar{T}_e}{\bar{x}^{1-4\alpha}}$

or $\eta = \frac{y}{x^\alpha}$; $\psi = f(\eta)x^{1-\alpha}$, $T - T_e = g(\eta)x^{1-4\alpha}$

$T = T_w$ = constant implies that $T_w - T_e = g(0)x^{1-4\alpha}$ = const. or $\alpha = \frac{1}{4}$

$\therefore \quad \eta = \frac{y}{x^{1/4}}$; $\psi = f(\eta)x^{3/4}$

For dimensional reasons, the above transformations can be written in a form similar to the Falkner-Skan transformation,

$$\eta = (\frac{u_c}{\nu L})^{1/2} y(\frac{L}{x})^{1/4}; \quad \psi = (u_c\nu L)^{1/2}(\frac{x}{L})^{3/4}f(\eta)$$

where u_c can be any velocity scale. An appropriate definition for it is $u_c = \sqrt{g\beta L(T_w - T_e)}$

9.4 For 2-D laminar natural convection flow over a vertical flat plate, the governing equations and their boundary conditions are

$$\frac{\partial u}{\partial x} + \frac{\partial v}{\partial y} = 0, \quad u\frac{\partial u}{\partial x} + v\frac{\partial u}{\partial y} = \nu\frac{\partial^2 u}{\partial y^2} + g\beta(T - T_e)$$

$$u\frac{\partial T}{\partial x} + v\frac{\partial T}{\partial y} = \frac{\nu}{Pr}\frac{\partial^2 T}{\partial y^2}$$

$$y = 0, \quad u = v = 0, \quad T = T_w \text{ or } \frac{\partial T}{\partial y} = -\frac{\dot{q}_w}{k}; \quad y = \delta, \ u = 0, \quad T = T_e$$

(a) For specified surface temperature use the transformations

$$\eta = \left\{\frac{g\beta(T_w - T_e)}{\nu^2 x}\right\}^{1/4} y, \quad \psi = \{g\beta(T_w - T_e)\nu^2 x^3\}^{1/4} f(\eta) \tag{5}$$

Then $u = \left(\frac{\partial\psi}{\partial y}\right)_x = \frac{\partial\psi}{\partial\eta}\frac{\partial\eta}{\partial y} = [g\beta(T_w - T_e)x]^{1/2} f'(\eta)$

$$v = -\left(\frac{\partial\psi}{\partial x}\right)_y = -\left(\frac{\partial\psi}{\partial x}\right)_\eta - \frac{\partial\psi}{\partial\eta}\frac{\partial\eta}{\partial x}$$

$$= -[g\beta(T_w - T_e)\nu^2 x^3]^{1/4}\left\{\frac{1}{4}(n + 3)\frac{f}{x} + f'\frac{\partial\eta}{\partial x}\right\}$$

$$\left(\frac{\partial u}{\partial x}\right)_y = \left(\frac{\partial u}{\partial x}\right)_\eta + \frac{\partial u}{\partial\eta}\frac{\partial\eta}{\partial x}$$

$$= [g\beta(T_w - T_e)x^2]^{1/2}\left\{\frac{f'}{x}\left(\frac{1}{2}n + \frac{1}{2}\right) + f''\frac{\partial\eta}{\partial x}\right\}$$

$$u\left(\frac{\partial u}{\partial x}\right)_y = \{g\beta(T_w - T_e)\}\left\{f'^2\left(\frac{1}{2} + \frac{1}{2}n\right) + f'f''\frac{\partial\eta}{\partial x}x\right\}$$

$$\frac{\partial u}{\partial y} = \frac{\partial u}{\partial\eta}\frac{\partial\eta}{\partial y} = \{g\beta(T_w - T_e)x\}^{1/2} f''\left\{\frac{g\beta(T_w - T_e)}{\nu^2 x}\right\}^{1/4}$$

$$v\frac{\partial u}{\partial y} = -\{g\beta(T_w - T_e)\}\left\{\frac{1}{4}(3 + n)ff'' + x\frac{\partial\eta}{\partial x}f'f''\right\}$$

$$\nu\frac{\partial^2 u}{\partial y^2} = \nu\{g\beta(T_w - T_e)x\}^{1/2} f'''\left\{\frac{g\beta(T_w - T_e)}{\nu^2 x}\right\}^{1/2} = \{g\beta(T_w - T_e)\}f'''$$

$$\left(\frac{\partial T}{\partial x}\right)_y = \left(\frac{\partial T}{\partial x}\right)_\eta + \frac{\partial T}{\partial\eta}\frac{\partial\eta}{\partial x} = (T_w - T_e)\left\{\frac{n}{x}\theta + \theta'\frac{\partial\eta}{\partial x}\right\}$$

$$u\frac{\partial T}{\partial x} = (T_w - T_e)\{g\beta(T_w - T_e)x\}^{1/2}\left\{\frac{n}{x}\theta + \theta'\frac{\partial\eta}{\partial x}\right\}f'$$

$$-v\frac{\partial T}{\partial y} = (T_w - T_e)\{g\beta(T_w - T_e)x\}^{1/2}\left\{\frac{1}{4}(3 + n)\frac{\theta'}{x}f + \theta'f'\frac{\partial\eta}{\partial x}\right\}$$

$$\frac{\nu}{Pr}\frac{\partial^2 T}{\partial y^2} = \frac{\theta''}{Pr}(T_w - T_e)\left\{\frac{g\beta(T_w - T_e)}{x}\right\}^{1/2}$$

Substituting the above equations into the momentum and energy eqs. and rearranging we get

$$f''' + \frac{3}{4} ff'' - \frac{1}{2}(f')^2 + \theta - \frac{1}{2} n [(f')^2 - \frac{1}{2} ff''] = 0 \qquad (9)$$

$$\frac{\theta''}{Pr} + \frac{3}{4} f\theta' + n(\frac{1}{4} f\theta' - f'\theta) = 0 \qquad (10)$$

$$\eta = 0, \quad f = f' = 0, \quad \theta = 1; \quad \eta = \eta_e, \quad f'_e = 0, \quad \theta = 0$$

$$\text{where } n = \frac{x}{T_w - T_e} \frac{d(T_w - T_e)}{dx}, \quad \theta(\eta) = \frac{T - T_e}{T_w - T_e}$$

(b) For specified wall flux, use

$$\eta = \{\frac{g\beta(T_w - T_e)}{\nu^2 x}\}^{1/4} y, \quad \psi = \{g\beta(T_w - T_e)\nu^2 x^3\}^{1/4} f(\eta)$$

$$T = T_e + T_e(\theta)\phi(x), \quad \phi(x) = -\frac{\dot{q}_w(x)}{k} \{\frac{\nu^2 x}{g\beta(T_w - T_e)}\}^{1/4}$$

Under this transformation, the momentum equation is the same as the equation in part (a) with θ there defined by $\theta = (T - T_e)/(T_w - T_e)$. To obtain the energy equation for this case,

$$(\frac{\partial T}{\partial x})_y = (\frac{\partial T}{\partial x})_\eta + \frac{\partial T}{\partial \eta}\frac{\partial \eta}{\partial x} = T_e\ \theta \frac{\partial \phi}{\partial x} + T_e\theta'\phi \frac{\partial \eta}{\partial x}$$

$$u\ (\frac{\partial T}{\partial x})_y = \{g\beta(T_w - T_e)x\}^{1/2} T_e\{\theta \frac{\partial \phi}{\partial x} f' + \phi\theta' f' \frac{\partial \eta}{\partial x}\}$$

$$\frac{\partial T}{\partial y} = \frac{\partial T}{\partial y}\frac{\partial \eta}{\partial y} = T_e\phi\theta' \{\frac{g\beta(T_w - T_e)}{\nu^2 x}\}^{1/4}$$

$$v\ \frac{\partial T}{\partial y} = -(T_e\phi\theta')\{g\beta(T_w - T_e)x\}^{1/2}\{(\frac{1}{4} n + \frac{3}{4})\frac{f}{x} + f'\frac{\partial \eta}{\partial x}\}$$

$$\frac{\nu}{Pr}\frac{\partial^2 T}{\partial y^2} = T_e\phi\theta''/Pr\ \{\frac{g\beta(T_w - T_e)}{x}\}^{1/2}$$

With the above relations, the energy equation becomes

$$+\frac{T_e\phi\theta''}{Pr}\{\frac{g\beta(T_w - T_e)}{x}\}^{1/2} = \{g\beta(T_w - T_e)x\}^{1/2}$$

$$(+T_e\phi)\ \{\frac{\theta}{\phi}\frac{\partial \phi}{\partial x} f' - (\frac{1}{4} n + \frac{3}{4})\frac{\theta' f}{x}\}$$

Dividing both sides by $(-T_e\phi)\{g\beta(T_w - T_e)/x\}^{1/2}$ and rearranging we can write the energy equation and its boundary condition as

$$\frac{\theta''}{Pr} + \frac{3}{4}\theta' f + \frac{1}{4} nf\theta' - \hat{n}\theta f' = 0$$

$$\eta = 0, \quad \theta' = 1; \quad \eta = \eta_e, \quad \theta = 0$$

9.5 $$u = \frac{\partial \psi}{\partial y} = [g\beta(T_w - T_e)\nu^2 x^3]^{1/4} \frac{\partial f}{\partial \eta}\frac{\partial \eta}{\partial y}$$

$$= [g\beta(T_w - T_e)\nu^2 x^3]^{1/4} [\frac{g\beta(T_w - T_e)}{\nu^2 x}]^{1/4} f'$$

$$= [g\beta(T_w - T_e)L]^{1/2} \left(\frac{x}{L}\right)^{1/2} f'$$

This implies that u is maximum when f' is maximum. From Fig. 9.2,

$f'_{max} = 0.55$ for Pr = 0.72, $\therefore$ $u_{max} = 0.55\left(\frac{x}{L}\right)^{1/2}[g\beta(T_w - T_e)L]^{1/2}$

9.6 From Fig. 9.2, the edge of the boundary layer is $\eta_e \simeq 5.5$. For

$T_w = 50°C$, $T_e = 15°C$, $T_m = 1/2(T_w + T_e) = 32.5°C$,

$\nu = 1.62 \times 10^{-5} m^2/s$, $\beta = 1/T_m$ and $x = 0.5m$,

$$y_e = \left\{\frac{\nu^2 x}{g\beta(T_w - T_e)}\right\}^{1/4} \eta_c = \left\{\frac{(1.62 \times 10^{-5})^2 \times 0.5}{9.8 \times 35/(273 + 32.5)}\right\}^{1/4} \times 5.5 = 0.018m$$

To avoid the interference of the layers, the plates must be separated at least two times y_e, i.e., 0.036m.

9.7 (a) Air: $T_w = 65°C$, $T_e = 15°C$, $T_f = \frac{1}{2}(T_w + T_e) = 40°C$, $x = 0.25m$

$\nu = 1.62 \times 10^{-5}\ m^2/s$ and Pr = 0.72 for air at 40°C

From Table 9.1, $F''_w = 0.9558$; $-\theta'_w = 0.3567$

$$Gr_x = \frac{g\beta x^3(T_w - T_e)}{\nu^2} = \frac{9.8 \times (0.25)^3 \times (65-15)}{(273+40) \times (1.65\times10^{-5})^2} = 8.985 \times 10^7$$

$$\therefore \quad c_f = \frac{2f''_w}{\sqrt{Gr_x}} = \frac{2 \times 0.9558}{\sqrt{8.985 \times 10^7}} = 2.02 \times 10^{-4}$$

$$Nu_x = -\theta'_w Gr_x^{1/4} = 0.3567 \times (8.985 \times 10^7)^{1/4} = 347$$

(b) Glycerin: $\nu = 2.2 \times 10^{-4}\ m^2/s$, $\beta = 5.24 \times 10^{-4}$,

Pr = 2.45 at 40°C

$$Gr_x = \frac{9.8 \times (0.25)^3 \times 5.24 \times 10^{-4}(65 - 15)}{(2.2 \times 10^{-4})^2} = 8.29 \times 10^4$$

$-\theta'_w = 0.55$, $f''_w = 0.77$ for Pr = 2.45

$$\therefore c_f = \frac{2 \times 0.77}{(8.29\times10^4)^{1/2}} = 5.35 \times 10^{-3},\ Nu_x = 0.55(8.29\times10^4)^{1/4} = 9.33$$

(c) Water: From Table B-2 $\beta = -\left[\frac{1}{\rho}\frac{\partial\rho}{\partial T}\right]_{T_f} = \frac{-1}{994.59} \times \frac{985.46 - 1000.52}{40}$

$= 3.75 \times 10^{-4}$, $\nu = 0.658 \times 10^{-6}$, Pr = 4.34 at 40°C

$$Gr_x = \frac{9.8 \times (0.25)^3 \times 3.75 \times 10^{-4}(65 - 15)}{(0.658 \times 10^{-6})^2} = 6.7 \times 10^9$$

$f''_w = 0.70$, $-\theta'_w = 0.66$ for Pr = 4.34 from Table 9.1

$$\therefore \quad c_f = \frac{2.0 \times 0.7}{(6.7 \times 10^9)^{1/2}} = 1.71 \times 10^{-5}$$

$$Nu_x = 0.66 \times (6.7 \times 10^9)^{1/4} = 1.89 \times 10^2$$

9.8 $T_f = 0.5(15° + 250) = 132°C = 405.5K$, $\rho = 0.8826\ kg/m^3$,

$c_p = 1.0140\ kJ/kgK$, $\mu = 2.288\ kg/ms$, $\nu = 25.90 \times 10^{-6}\ m^2/s$,

$k = 0.03365\ W/mK$, $Pr = 0.689$, $\kappa = 0.3760 \times 10^{-4}\ m^2/s$

(a) Semi-Infinite: From Eq. (9.60) $Nu = \frac{\dot{q}_w}{(T_w - T_e)} \frac{L}{k} = - \frac{Gr_L^{1/5}}{\xi^{2/5}} H'(0)$

where $H'(0) = 0.354$ at $Pr = 0.689$, $\xi = x/L$

$$\dot{q}_w = - (T_w - T_e) \frac{k}{L} \frac{Gr_L^{1/5}}{\xi^{2/5}} H'(0)$$

$$\bar{\dot{q}}_w = \frac{1}{L} \int_0^L \dot{q}_w dx = -(T_w - T_e) \frac{k}{L} Gr_L^{1/5} H'(0) \int_0^1 \frac{1}{\xi^{2/5}} d\xi$$

$$= -(T_w - T_e) k Gr_L^{1/5} \frac{5}{3} H'(0)/L = \frac{(T_w - T_e)k}{L} Gr_L^{1/5} \times 0.59$$

(b) Finite: $\bar{\dot{q}}_w = \frac{-k(T_w - T_e)}{L} 0.28(Gr_L Pr)^{1/3}$

$$= - \frac{k(T_w - T_e)}{L} Gr_L^{1/3} \times 0.25$$

$$Nu_x = \frac{\hat{h}x}{k} = \frac{\dot{q}_w}{(T_w - T_e)} \frac{x}{k}$$

$$\overline{Nu} = \int_0^L Nu_x dx = \frac{-\dot{q}_w}{(T_w - T_e)} \int_0^L \frac{x}{k} dx = \frac{L^2}{2k} \frac{\dot{q}_w}{(T_w - T_e)}$$

9.9 (a) c_f, and Nu are computed from

$$c_f = \frac{2f''_w}{\sqrt{\xi}} \frac{\sqrt{Gr_L}}{R_L^{3/2}} = \frac{2f''_w}{\sqrt{R_x}}, \quad Nu_x = \frac{-\theta'_w}{\sqrt{\xi}} \frac{R_L^{3/2}}{\sqrt{Gr_L}} = -\theta'_w \sqrt{R_x}$$

where $\xi = \frac{x}{L} Ri$, $Ri = \frac{u_c^2}{u_e^2} = \frac{g\beta L(T_w - T_e)}{u_e^2}$, $R_x = \frac{u_e x}{\nu}$

and f''_w and θ'_w are dependent on ξ. For $u_e = 10\ m/s$,

$L = 1.5m$, $T_w = 120°C$, $T_e = 20°C$, $T_f = \frac{1}{2}(T_w + T_e) = 70°C$

$\nu = 2.076 \times 10^{-5}\ m^2/s$ for air at 70°C

$$Ri = \frac{9.8 \times 1.5 \times (120 - 20)/343}{10^2} = 0.04286$$

The values of c_f and Nu_x at x = 0.5m, 1.0m and 1.5m follow from their definitions and from Figs. 9.10 and 9.11 and are tabulated below

x	ξ	$-\theta'_w$	Nu_x	f''_w	$c_f \times 10^3$
0.5	0.0143	0.296	145.3	0.34	1.385
1.0	0.0286	0.298	206.8	0.36	1.037
1.5	0.0429	0.30	255.0	0.38	0.089

As shown in Figs. (9.10) and (9.11), while f''_w and $-\theta'_w$ increase as ξ increases for a heated vertical surface, they are constant (and equal to 0.332 and 0.2957 for Pr = 0.72) for a heated horizontal surface. As a result, c_f and Nu_x are higher for a vertical plate than for a horizontal one.

(b) The total heat flux, $Q_w = \int_0^L w\dot{q}_w dx$ is related to Nu by

$$\dot{q}_w = (T_w - T_e)\, Nu_x \frac{k}{x}$$

$$\therefore Q_w = w(T_w - T_e)k \int_0^L -\theta'_w \sqrt{R_x}/x\, dx = w(T_w - T_e)k\, (-\overline{\theta'_w})\, 2\sqrt{R_L}$$

$$= 0.5(120 - 20) \times 0.03 \times 0.298 \times 2\left[\frac{10 \times 1.5}{2.076 \times 10^{-5}}\right]^{1/2} = 760W$$

9.10 For an ideal gas, $Ri \equiv \frac{gL}{u_e^2} \frac{(T_w - T_e)}{T}$

(a) $Ri = \frac{9.81 \times 0.01}{0.01} \times \frac{20}{300} = 0.654$

(b) $Ri = \frac{9.81 \times 0.01}{1} \times \frac{20}{300} = 6.54 \times 10^{-3}$

(c) $Ri = \frac{9.81 \times 0.01}{1} \times \frac{700}{300} = 0.229$: Analysis is only valid for $\frac{\Delta T}{T} \to 0$.

9.11 (a) $Ri = \frac{9.81 \times 0.01}{0.01} \times \frac{200}{300} = 6.54$

(b) $Re = \frac{0.1 \times 0.01}{1.6} \times 10^5 = 62.5$

$z - z_0 = 10$, $f''_w \simeq 20$, $-\theta_w = 1.0$

$$c_f/2 = \left(\frac{6.54}{62.5}\right)^{0.5} \frac{20}{(0.1/0.01)^{0.5}} = 2.03$$

$$Nu = 1 \times (10)^{0.5} \times \left(\frac{62.5}{6.54}\right)^{0.5} = 9.78$$

9.12 $f''_w \simeq 20 \times 1.1$, $-\theta'_w = 10 \times 5/6$

$\therefore$ $c_f/2 = 2.03 \times 1.1 = 2.2$, $Nu = 9.78 \times 5/6 \simeq 8.15$

PART 2

COMPUTER PROGRAMS

Introductory Remarks

Sections 13.3 and 13.5 and 14.1 to 14.4 of the book entitled "Physical and Computational Aspects of Convective Heat Transfor" provide listings of computer programs which allow the solution of the equations representing conservation of mass, momentum and energy for two internal and four external boundary-layer flows. The following six sections repeat the descriptions of these computer programs and provide additional comments to facilitate their use. The first section considers a duct flow problem in which the velocity field is fully developed so that only the energy equation has to be solved. This introductory problem is followed by a description of the Fortran program for the solution of the equations for two-dimensional coupled wall boundary layers for both laminar and turbulent flows and the remaining four problems are solved with the aid of versions of this general program.

For the convenience of the user, the complete listings of the computer programs of Sections 1.1 to 1.6 are included in one DOS 5-1/4" floppy diskette which can be obtained from the author. It is formatted for IBM PC/XT or compatible computers. The programs and data files are labelled as FILENAME.SPEC. Here FILEMANE designates chapter and problem, i.e. PR428 for Problem 4.28 and SPEC designates either the Fortran listing (FOR) or text input file (DAT). The Fortran codes use FORTRAN files in which 5 indicates input data (read), 6 output data (write) and 9 creates an output file suitable for machine plotting (write). The computer programs also contain input and output information which allows the user to check his or her ability to use the program.

1.1 Computer Program for Pipe Flow with Heat Transfer

This introductory problem assumes that the velocity field of the flow in a circular pipe of uniform cross-section is fully developed with a uniform wall temperature different from that of the fully-developed flow at the beginning of the thermal boundary layer. It requires the solution of the energy equation in the form

$$u \frac{\partial T}{\partial x} = \frac{\nu}{Pr} \frac{1}{r} \frac{\partial}{\partial y} [r(1 + \frac{Pr}{Pr_t} \varepsilon_m^+) \frac{\partial T}{\partial y}]$$

Laminar flow solutions can be obtained by setting ε_m^+ to zero and altering the shape of the fully developed velocity profile.

The problem consists of a MAIN program and three subroutines: COEF, SOLV2 and OUTPUT. The following sections present a brief description of each routine. They should be read with the relevant listing printed out from the diskette, which accompanies this manual.

1.1.1 MAIN

The MAIN routine contains the overall logic of computations, specifies the initial temperature profiles at $x = x_0$, generates a nonuniform grid across the duct and accounts for the boundary-layer growth. The grid has the property that the ratio of lengths of any two adjacent intervals is a constant; that is, $\eta_j = \eta_{j-1} + h_j$, where $h_j = Kh_{j-1}$. The distance to the j-th line is given by the following formula*:

$$\eta_j = h_1 \frac{K^j - 1}{K - 1}, \qquad j = 1, 2, 3, \ldots, J, \qquad K > 1 \qquad (13.48)$$

There are two parameters: h_1, the length of the first $\Delta\eta$ step, and K, the ratio of two successive steps. The total number of points, J, is calculated by the following formula:

$$J = \frac{\ln[1 + (K - 1)(\eta_e/h_1)]}{\ln K} \qquad (13.49)$$

For laminar flows, we generally take $K(\equiv VGP) = 1$, so that the grid is uniform and with the choice of η_e ($\equiv$ ETAE) = 8 and h_1 ($\equiv$ DETA(1)] = 0.2, we take 41 grid points across the layers, which is sufficient for most laminar-flow calculations. For turbulent flows, we usually select the parameters h_1 and K and calculate η_e. An idea of the number of points taken across the shear layer with these parameters for different η_e-values can be obtained from the following figure. For example, for $h_1 = 0.025$ and $K = 1.15$ and $\eta_e = 10$, the ratio of η_e/h_1 is 400 and the initial number of points is approximately 30 points. Since the boundary-layer thickness

*Unless otherwise stated, the equation numbers follow those in the book.

grows for turbulent flows, at later x stations we may have around 40-50 grid points.

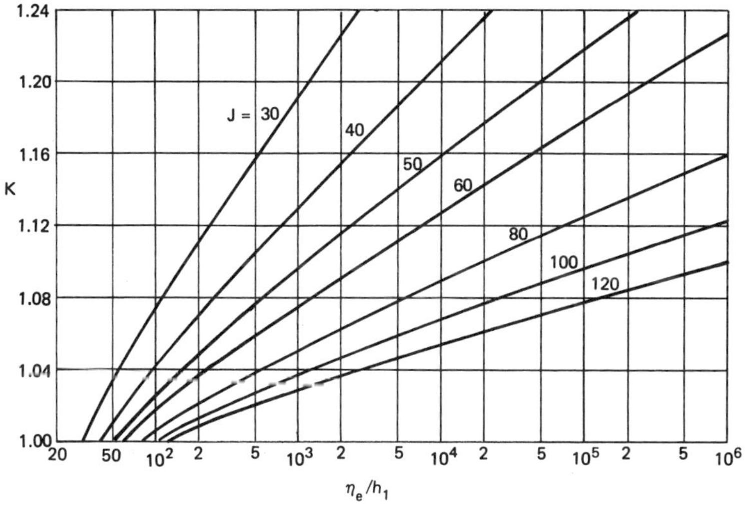

Variation of K with h_1 for different η_e-values.

This routine also contains the input data that consist of

- NXT: total number of x stations
- ITURB: 0 for laminar and 1 for turbulent flow
- IWBCOE: 1 for specified wall temperature and 0 for specified wall heat flux
- IEBCOE: 1 for specified edge temperature and 0 for zero edge-temperature gradient

and ETAE, DETA(1), VGP, PR, GWA, REY. In the present problem IWBCOE = 1, IEBCOE = 1, and GWA corresponds to the parameter needed in the smoothing function, Eq. (4.90), used for the wall temperature. We use this equation (see subroutine COEF) in order to avoid the difficulties associated with the jump in the wall temperature as was described in Section 4.3. The parameter REY is the Reynolds number based on radius, $u_0 r_0/\nu$ ($\equiv R_d/2$). Note that both PR and REY are needed only for turbulent flows, since for laminar flow they do not appear in the dimensionless form of the equations.

1.1.2 Subroutine COEF

Subroutine COEF defines the coefficients of the finite-difference form of the energy equation given by Eqs. (13.37). They are valid for both cases in which the energy equation is expressed either in transformed, SW = 0.0, or physical, SW = 1.0, variables. For turbulent flows the coefficients remain unchanged except that a_1 takes the form given by Eq. (7.31). The subroutine also contains expressions for calculating the velocity profiles for laminar and turbulent flows. In the former case, the profile is computed from Eq. (5.20), u = 1 - (r), and in the latter case from Eq.(7.13), with friction factor computed by Eq. (7.17). The ε_m^+ and Pr_t expressions are obtained from Eqs. (7.9), (7.10) and (6.26), respectively. Note that at y = 0 the latter becomes $Pr_t = (\kappa/\kappa_h)(B/A)$.

1.1.3 Subroutine OUTPUT

Subroutine OUTPUT prints out the g, p, u profiles across the shear layer and shifts them prior to the calculations at a new x-station. It also computes the dimensionless mixed-mean temperature given by Eq. (5.4) and the Nusselt number given by Eq. (5.23), which in terms of dimensionless temperature and temperature gradient g_w' is $2g_w'/g_m$ [see Eq. (P5.10)]. In the region before the thermal shear layers merge, the dimensionless bulk temperature g_m is computed according to Eq. (P5.8) and g_w' is computed from (P5.11). After the thermal shear layers merge, they are computed from Eqs. (P5.9) and $(\partial g/\partial y)_w$, respectively. This subroutine also determines the x-station (denoted by SW) where the calculations switch to physical variables as discussed in Section 13.3.

1.1.4 Subroutine SOLV2

Subroutine SOLV2 contains the algorithm used to solve the linear system (13.36) and (13.38) which is written as

$$A\,\underset{\sim}{\delta} = \underset{\sim}{r} \qquad (13.28)$$

with the block-elimination method discussed in Section 13.2. The notation used in the algorithm closely follows that used in Section 13.2 with $\underset{\sim}{\delta}_j$ and $\underset{\sim}{r}_j$ defined by

$$\underset{\sim}{\delta}_j = \begin{vmatrix} T_j \\ p_j \end{vmatrix} \qquad \underset{\sim}{r}_j = \begin{vmatrix} (r_1)_j \\ (r_2)_j \end{vmatrix} \qquad (13.30a)$$

and the 2 x 2 matrices A_j, B_j, C_j defined by

$$A_0 \equiv \begin{vmatrix} \alpha_0 & \alpha_1 \\ -1 & \frac{-h_1}{2} \end{vmatrix}, \qquad A_j \equiv \begin{vmatrix} (s_3)_j & (s_1)_j \\ -1 & \frac{-h_{j+1}}{2} \end{vmatrix} \qquad 1 \le j \le J-1$$

$$A_J \equiv \begin{vmatrix} (s_3)_J & (s_1)_J \\ \beta_0 & \beta_1 \end{vmatrix}, \qquad B_j \equiv \begin{vmatrix} (s_3)_j & (s_2)_j \\ 0 & 0 \end{vmatrix} \qquad 1 \le j \le J$$

$$C_j \equiv \begin{vmatrix} 0 & 0 \\ 1 & \frac{-h_{j+1}}{2} \end{vmatrix} \qquad 0 \le j \le J-1 \qquad (13.30b)$$

Here α_0, α_1 and β_0, β_1 represent the parameters used to write the wall and edge boundary conditions, respectively. For the case of specified wall temperature, we set $\alpha_0 = 1$ and $\alpha_1 = 0$ and for specified wall heat flux $\alpha_0 = 0$ and $\alpha_1 = 1$. For the case of specified edge temperature, we set $\beta_0 = 1$, $\beta_1 = 0$ and $\beta_0 = 0$, $\beta_1 = 1$ when the edge temperature gradient is specified. See also Subroutine INPUT.

1.1.5 Sample Calculations

To illustrate the use of this code, we now present results for a laminar flow in a circular pipe with constant wall temperature obtained by starting the calculations at $\hat{x}_0 = 1 \times 10^{-6}$ with $\eta_e = 8$, $h_1 = 0.20$, $K = 1$ and $a = 2 \times 10^6$ [see Eq. (4.90)] and with Nusselt number computed from Eq. (5.23). In the region where g_w changes according to Eq. (4.90), 15 x stations were taken. For the accuracy of the numerical solutions, see the discussion of pp. 405 and 406, and for the application of this code to both laminar and turbulent flows, see the problems in Chapters 5 and 7, respectively.

Results: Nusselt number variation with nondimensional distance for fully-developed pipe flow with constant wall temmperature.

x^+/Pe	Nu	x^+/Pe	Nu
1×10^{-4}	28.3	6×10^{-2}	3.89
5×10^{-4}	16.5	1×10^{-1}	3.71
1×10^{-3}	12.9	3×10^{-1}	3.63
4×10^{-3}	8.02	∞	3.63
2×10^{-2}	4.92		

1.2 Computer Program for Coupled Wall Boundary-Layer Flows

This general program allows the solution of steady, two-dimensional laminar and turbulent boundary-layer flows and provides the basis upon which the programs of Sections 1.3 to 1.6 are founded. It consists of seven subroutines MAIN, INPUT, IVPL, EDDY, COEF, OUTPUT and SOLV5 and these are described below. The flow diagram provided in Fig. 13.8 is also presented here for completeness.

1.2.1 MAIN

Subroutine MAIN contains the overall logic of the computations, generates the grid normal to the flow, calculates the fluid properties, and accounts for the boundary-layer growth. It also checks the convergence of the iterations by using the wall shear parameter v_0 as the convergence criterion. For laminar flows, calculations are stopped when

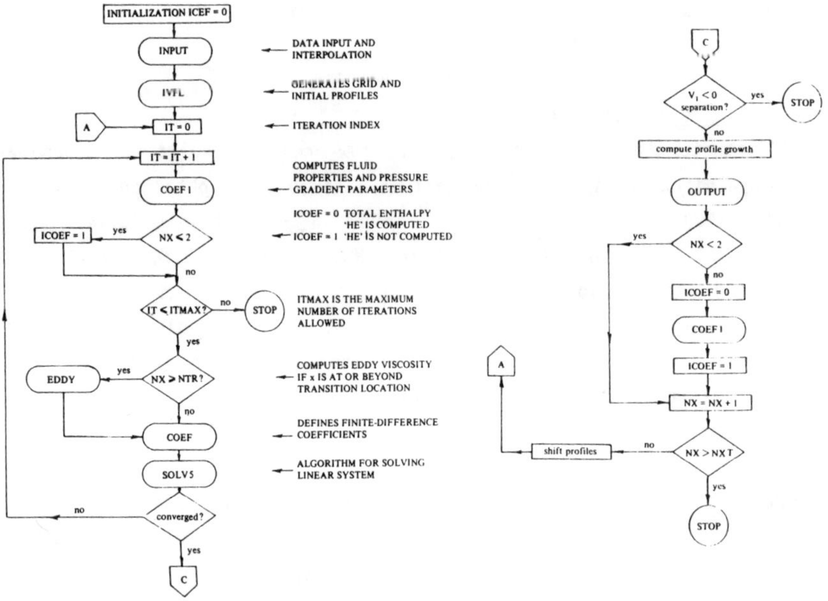

Fig. 13.8 Flow diagram for the computer program of Section 13.5.

$$\left|\delta v_0^{(i)}\right| < 10^{-5}, \tag{13.69}$$

which gives about four-figure accuracy, sufficient for most predicted quantities. If v_0 becomes negative during any iteration, the calculations are stopped. For turbulent flows, the above equation is expressed on a percentage basis and is written as

$$\left|\frac{\delta v_0^{(i)}}{v_0 + \delta v_0^{(i)}/2}\right| < 0.02 \tag{13.70}$$

The grid generated across the layer is identical to that described in Section 13.3, with typical values of h_1 and K equal to 0.2 and 1, respectively, for laminar flows and 0.01 and 1.14, respectively, for turbulent flows. For a flow consisting of both laminar and turbulent flows, they may be chosen as 0.01 and 1.14.

The parameters μ, c, C, b, e and d appearing in the momentum and energy equations are functions of static temperature. In this routine c, C, b, e and d are defined by Eq. (10.7) for laminar flows and by Eq. (11.47) for turbulent flows. The dynamic viscosity μ is computed from Sutherland's formula, Eq. (1.4a), once the total-enthalpy ratio g is computed from the energy equation.

For most laminar boundary-layer flows the transformed boundary-layer thickness $\eta_e(x)$ is almost constant and a value of $\eta_e = 8$ is sufficient for the velocity to attain 0.9999 of its external-stream value. However, for turbulent boundary layers, $\eta_e(x)$ generally increases with increasing x and to estimate $\eta_e(x)$ we always require that $\eta_e(x_n) \geq \eta_e(x_{n-1})$. Then in subroutine MAIN, when the computations on $x = x_n$ (for any $n \geq 1$) have been completed, we test to see if $|v_J^n| \leq \epsilon_v$, where $\eta_J = \eta_e(x_n)$. Otherwise, we set $J_{new} = J_{old} + t$, where t is a number of added points, say t = 3. In this case we also specify values of f_j^n, u_j^n, v_j^n, g_j^n, p_j^n, b_j^n, and e_j^n for the new η_j points. We take the values of $u_j = 1$, $g_j = 1$, $v_j^n = v_J^n$, $p_j^n = p_J^n$, $c_j^n = c_J^n$, $d_j^n = d_J^n$, $e_j^n = e_J^n$, $\mu_j^n = \mu_J^n$, $C_j^n = C_J^n$, $f_j^n = (\eta_j - \eta_e)u_J^n + f_J^n$. This is also done for the same parameters of the previous station (n - 1).

Fortran Name	Symbol
ITMAX	Iteration count
CEL,P1P,P2P	α_n, $m_1^n + \alpha_n$, $m_2^n + \alpha_n$, respectively
DELV(1)	δv_0^n
NP	J

1.2.2 Subroutine INPUT

Subroutine INPUT specifies the wall boundary conditions for the energy equation, the total number of x stations (NXT), the transition location (NTR), the dimensionless pressure gradient m_2 at the first x-station P2(1), and the variable grid parameters h_1 and K. In addition, we specify the freestream parameters M_∞, T_∞, p_∞ (RMI, TI, PI, respectively) and the molecular Prandtl number Pr as well as u_e/u_∞ as a function of surface distance x. For all stations except the first and last, the derivative du_e/dx, appearing in m_1 and m_2, is obtained from three-point Lagrange interpolation formulas.

$$\left(\frac{du_e}{dx}\right)_n = -\frac{u_e^{n-1}}{A_1}(x_{n+1} - x_n) + \frac{u_e^n}{A_2}(x_{n+1} - 2x_n + x_{n-1}) + \frac{u_e^{n+1}}{A_3}(x_n - x_{n-1}) \tag{13.71}$$

where

$$A_1 = (x_n - x_{n-1})(x_{n+1} - x_{n-1}), \quad A_2 = (x_n - x_{n-1})(x_{n+1} - x_n),$$

$$A_3 = (x_{n+1} - x_n)(x_{n+1} - x_{n-1})$$

The derivative du_e/dx at the endpoint $n = N$ is computed from

$$\left(\frac{du_e}{dx}\right)_N = \frac{u_e^{N-2}}{A_1}(x_N - x_{N-1}) - \frac{u_e^{N-1}}{A_2}(x_N - x_{N-2}) + \frac{u_e^N}{A_3}(2x_N - x_{N-2} - x_{N-1}), \tag{13.72}$$

where now

$$A_1 = (x_{N-1} - x_{N-2})(x_N - x_{N-2}), \quad A_2 = (x_{N-1} - x_{N-2})(x_N - x_{N-1}),$$

$$A_3 = (x_N - x_{N-1})(x_N - x_{N-2})$$

The freestream values μ_∞, u_∞, ρ_∞, and H_e are calculated for air from

$$\mu_\infty = 1.45 \times 10^{-6} \frac{T_\infty^{3/2}}{T_\infty + 100} \text{ kg/m s}, \quad u_\infty = 20.04 M_\infty \sqrt{T_\infty} \text{ m/s}$$

$$\rho_\infty = \frac{p_\infty}{287 T_\infty} \text{ kg/m}^3, \quad H_e = 1004 T_\infty + \frac{1}{2} u_\infty^2 \text{ m}^2/\text{s}^2 \tag{13.73}$$

where the temperature is expressed in kelvin (K). The edge values of T_e and p_e are computed from

$$\frac{T_e}{T_\infty} = 1 - \frac{\gamma - 1}{2} M_\infty^2 \left[\left(\frac{u_e}{u_\infty}\right)^2 - 1\right], \quad \frac{p_e}{p_\infty} = \left(\frac{T_e}{T_\infty}\right)^{\gamma/\gamma-1} \tag{13.74}$$

The edge values of μ and ρ are computed from formulas identical to those given by Eq. (13.73) except that freestream values of temperature and pressure are replaced by their edge values.

The boundary conditions for the energy equation are written so that its solution can be obtained either for specified wall temperature by reading α_0 = ALFA0 = 1.0, α_1 = ALFA1 = 0.0 or for specified heat flux by reading $\alpha_0 = 0.0$, $\alpha_1 = 1.0$. In either case the dimensionless wall temperature g_w or temperature gradient p_w are read through WW(I). Note that the read-in values of UE(I) are in dimensional form (m/s).

1.2.3 Subroutine IVPL

Subroutine IVPL is used to generate the initial velocity profiles for a compressible similar laminar flow. At first, however, the calculations are done for a similar incompressible flow in order to generate better profiles for the compressible flow by iteration. The dimensionless total enthalpy ratio g_j is taken as unity, and its derivative p_j is set equal to zero. The initial profiles of f_j, u_j and v_j are computed from the velocity profile given by Eq. (4.59) and from the expressions based on this equation,

$$f = \frac{\eta_e}{4}\left(\frac{\eta_j}{\eta_e}\right)^2\left[3 - \frac{1}{2}\left(\frac{\eta_j}{\eta_e}\right)^2\right], \quad u_j = \frac{3}{2}\frac{\eta_j}{\eta_e} - \frac{1}{2}\left(\frac{\eta_j}{\eta_e}\right)^3,$$

$$v_j = \frac{3}{2}\frac{1}{\eta_e}\left[1 - \left(\frac{\eta_j}{\eta_e}\right)^2\right]$$

1.2.4 Subroutine EDDY

Subroutine EDDY contains the formulas used in the Cebeci-Smith eddy-viscosity formulation and for simplicity we have excluded the low-Reynolds-number and the mass-transfer effects. These capabilities, if desired, can easily be incorporated. With subscripts i and o denoting inner and outer regions, the eddy-viscosity formulas are given by

$$(\varepsilon_m)_i = L^2 \frac{\partial u}{\partial y}\gamma_{tr}\gamma, \quad (\varepsilon_m)_i \le (\varepsilon_m)_o \tag{13.75a}$$

$$(\varepsilon_m)_o = 0.0168 \left|\int_0^\infty (u_e - u)dy\right| \gamma_{tr}\gamma, \quad (\varepsilon_m)_o \ge (\varepsilon_m)_i, \tag{13.75b}$$

where

$$L = 0.4y[1 - \exp(-\frac{y}{A})], \quad A = 26\left(\frac{\rho}{\rho_w}\right)^{1/2}\frac{\nu_e}{N}u_\tau^{-1},$$

$$N^2 = 1 - 11.8\frac{\mu_w}{\mu_e}\left(\frac{\rho_e}{\rho_w}\right)^2 p^+, \quad u_\tau = \left(\frac{\tau_w}{\rho_w}\right)^{1/2}, \quad p^+ = \frac{\nu_e u_e}{u_\tau^3}\frac{du_e}{dx} \tag{13.76}$$

$$\gamma_{tr} = 1 - \exp\left[- G_{tr}(x - x_{tr})\int_{x_{tr}}^{x}\frac{dx}{u_e}\right], \quad \gamma = [1 + 5.5\left(\frac{y}{\delta}\right)^6]^{-1}$$

$$G_{tr} = 8.33 \times 10^{-4}\frac{u_e^3}{\nu_3^2}(R_x)^{-1.34}, \quad R_x = \frac{u_e x}{\nu_e}$$

In terms of transformed variables defined in Chapter 13 these formulas become (noting that v = f")

$$(\varepsilon_m^+)_i = \frac{0.16}{c^2}\frac{\mu_e}{\mu}\sqrt{R_x}\, I^2 v^2 [1 - \exp(-\frac{y}{A})]^2 \gamma_{tr}\gamma \tag{13.77a}$$

$$(\varepsilon_m^+)_o = \frac{0.0168}{c}\frac{\mu_e}{\mu}\sqrt{R_x}\left[\int_0^{\eta_e} c(1 - u)d\eta\right]\gamma_{tr}\gamma, \tag{13.77b}$$

where with $\varepsilon_m^+ = \frac{\varepsilon_m}{\nu}$, $I = \int_0^\eta c\,d\eta$, $\frac{y}{A} = \frac{N}{26}c^{-3/2}\frac{C_w}{C}R_x^{1/4} I v_w^{1/2}$

$$N^2 = 1 - 11.8\frac{\mu_w}{\mu_e}c_w^2 p^+, \quad p^+ = \frac{m_2}{R_x^{1/4}}\left(\frac{\mu_e}{\mu_w}\right)^{3/2}\frac{1}{v_w^{3/2}}$$

1.2.5 Subroutine COEF

Subroutine COEF contains the coefficients of the linearized momentum and energy equations written in the form given by Eq. (13.62). The Fortran notation for some of the typical parameters are given in the accompanying table.

Fortran Name	Symbol
FB,UB,GB,CB	$f^n_{j-1/2}$, $u^n_{j-1/2}$, $g^n_{j-1/2}$, $c^n_{j-1/2}$
USB,FVB,FPB	$(u^2)^n_{j-1/2}$, $(fv)^n_{j-1/2}$, $(fp)^n_{j-1/2}$
DERDV	$[(bv)^n_j - (bv)^n_{j-1}]h^{-1}_j$
CFB,CUB,CGB,CCB	$f^{n-1}_{j-1/2}$, $u^{n-1}_{j-1/2}$, $g^{n-1}_{j-1/2}$, $c^{n-1}_{j-1/2}$
CDEREP	$h^{-1}_j[(ep)^{n-1}_j - (ep)^{n-1}_{j-1}]$
CRB,CLB	$R^{n-1}_{j-1/2}$, $L^{n-1}_{j-1/2}$ [see Eqs. (13.58a) and (13.58b)]
CTB,CMB	$T^{n-1}_{j-1/2}$, $M^{n-1}_{j-1/2}$ [see Eqs. (13.59a) and (13.59b)]
S1(J) to S8(J)	$(s_1)_j$ to $(s_8)_j$ [see Eq. (13.64)]
B1(J) to B10(J)	$(\beta_1)_j$ to $(\beta_{10})_j$ [see Eq. (13.65)]
R(1,J) to R(5,J)	$(r_1)_j$ to $(r_5)_j$ [see Eq. (13.63)]

1.2.6 Subroutine OUTPUT

Subroutine OUTPUT prints out the desired profiles such as f_j, u_j, v_j, g_j, p_j and b_j as functions of η_j and shifts them prior to the calculations at a new x-station. It also computes the boundary-layer parameters θ, δ^*, H, c_f, Nu_x, St_x, R_θ, R_{δ^*} and R_x whose definitions are summarized below for completeness.

$$\theta = \int_0^\infty \frac{\rho u}{\rho_e u_e}\left(1 - \frac{u}{u_e}\right)dy = \frac{x}{\sqrt{R_x}} \int_0^\infty f'(1 - f')d\eta \tag{13.78a}$$

$$\delta^* = \int_0^\infty \left(1 - \frac{\rho u}{\rho_e u_e}\right)dy = \frac{x}{\sqrt{R_x}} \int_0^\infty (c - f')d\eta \tag{13.78b}$$

$$H = \frac{\delta^*}{\theta}, \qquad c_f = \frac{\tau_w}{(1/2)\rho_e u_e^2} = \frac{2C_w}{\sqrt{R_x}} f''_w \tag{13.78c,d}$$

$$Nu_x = \frac{\dot{q}_w x}{(T_w - T_e)k} = \frac{C_w g'_w \sqrt{R_x}}{1 - g_w} \tag{13.78e}$$

$$St_x = \frac{\dot{q}_w}{\rho_e u_e(H_w - H_e)} = \frac{C_w g'_w}{Pr\,\sqrt{R_x}\,(1 - g_w)} \tag{13.78f}$$

$$R_\theta = \frac{u_e\theta}{\nu_e}\ , \quad R_{\delta*} = \frac{u_e\delta^*}{\nu_e}, \quad R_x = \frac{u_e x}{\nu_e} \tag{13.78g}$$

1.2.7 Subroutine SOLV5

Subroutine SOLV5 contains the algorithm used to solve the linear system (13.62) and (13.66) which is written in the same form as (13.28) with the block-elimination method. In this case $\underset{\sim}{\delta}_j$ and $\underset{\sim}{r}_j$ are defined by

$$\underset{\sim}{\delta}_j \equiv \begin{vmatrix} \delta f_j \\ \delta u_j \\ \delta v_j \\ \delta g_j \\ \delta p_j \end{vmatrix} \quad 0 \le j \le J, \qquad \underset{\sim}{r}_0 \equiv \begin{vmatrix} 0 \\ 0 \\ 0 \\ (r_4)_1 \\ (r_5)_1 \end{vmatrix} \tag{13.67a}$$

$$\underset{\sim}{r}_j \equiv \begin{vmatrix} (r_1)_j \\ (r_2)_j \\ (r_3)_j \\ (r_4)_j \\ (r_5)_j \end{vmatrix} \quad 1 \le j \le J-1, \qquad \underset{\sim}{r}_J \equiv \begin{vmatrix} (r_1)_J \\ (r_2)_J \\ (r_3)_J \\ 0 \\ 0 \end{vmatrix} \tag{13.67b}$$

and the 5 x 5 matrices A_j, B_j, C_j by

$$A_0 \equiv \begin{vmatrix} 1 & 0 & 0 & 0 & 0 \\ 0 & 1 & 0 & 0 & 0 \\ 0 & 0 & 0 & \alpha_0 & \alpha_1 \\ 0 & -1 & -\frac{h_1}{2} & 0 & 0 \\ 0 & 0 & 0 & -1 & -\frac{h_1}{2} \end{vmatrix}$$

$$A_J \equiv \begin{vmatrix} 1 & -\frac{h_J}{2} & 0 & 0 & 0 \\ (s_3)_J & (s_5)_J & (s_1)_J & (s_7)_J & 0 \\ (\beta_3)_J & (\beta_5)_J & (\beta_9)_J & (\beta_7)_J & (\beta_1)_J \\ 1 & 1 & 0 & 0 & 0 \\ 0 & 0 & 0 & 1 & 0 \end{vmatrix} \tag{13.68a}$$

$$A_j \equiv \begin{vmatrix} 1 & -\frac{h_j}{2} & 0 & 0 & 0 \\ (s_3)_j & (s_5)_j & (s_1)_j & (s_7)_j & 0 \\ (\beta_3)_j & (\beta_5)_j & (\beta_9)_j & (\beta_7)_j & (\beta_1)_j \\ 0 & -1 & -\frac{h_{j+1}}{2} & 0 & 0 \\ 0 & 0 & 0 & -1 & -\frac{h_{j+1}}{2} \end{vmatrix} \quad 1 \le j \le J-1 \tag{13.68b}$$

$$B_j \equiv \begin{vmatrix} -1 & -\frac{h_j}{2} & 0 & 0 & 0 \\ (s_4)_j & (s_6)_j & (s_2)_j & (s_8)_j & 0 \\ (\beta_4)_j & (\beta_6)_j & (\beta_{10})_j & (\beta_8)_j & (\beta_2)_j \\ 0 & 0 & 0 & 0 & 0 \\ 0 & 0 & 0 & 0 & 0 \end{vmatrix}, \quad 1 \le j \le J \tag{13.68c}$$

$$C_j \equiv \begin{vmatrix} 0 & 0 & 0 & 0 & 0 \\ 0 & 0 & 0 & 0 & 0 \\ 0 & 0 & 0 & 0 & 0 \\ 0 & 1 & -\frac{h_{j-1}}{2} & 0 & 0 \\ 0 & 0 & 0 & 0 & -\frac{h_{j+1}}{2} \end{vmatrix}, \quad 0 \le j \le J-1 \tag{13.68d}$$

See subroutine INPUT for the definitions of α_0 and α_1 used to specify the wall boundary conditions for the energy equation.

1.3 Computer Program for Forced and Free Convection Between Two Vertical Parallel Plates

The problem considered here involves laminar developing flow, a buoyancy force and two surfaces at the same uniform temperature different from that of the initial flow. It requires the solution of the equations given by (9.79) to (9.82), which can be obtained with the computer program of Section 1.2 with modifications to COEF, IVPL and MAIN as indicated below and to subroutine SOLV5 whose statement 40 should be preceded by the additional statement:

IF(IPROB.EQ.2)RETURN. (14.17)

Here IPROB is a flag which specifies whether the equations are solved for the standard problem (IPROB=1) or for the variational problem (IPROB=2), as discussed in Section 14.1.

For this problem, the IVPL subroutine of Section 1.2.3 is used to define the initial velocity profiles without modifications and by setting $b_j = 1$, $d_j = 0$, $e_j = 1/Pr$ and $c_j = 1$. This subroutine is used to calculate the initial temperature profile g and its derivative p, from

$$g = 1 - \frac{\eta}{\eta_e}, \qquad p = -\frac{1}{\eta_e} \tag{14.18}$$

The COEF subroutine of Section 1.2.5 is also used with some minor modifications and with P2(NX) set equal to zero and P1(NX) = 0.5 in MAIN and with the same coefficients $(s_k)_j$ (k = 1, ..., 6) and $(\beta_\ell)_j$. The coefficients $(s_7)_j$ and $(s_8)_j$ are defined by new expressions and $(r_k)_j$ are modified for standard and variational problems as indicated in the revised COEF subroutine.

The revised MAIN contains the INPUT, GRID, and OUTPUT subroutines and the logic of the calculations. It also contains Eqs. (14.9) and (14.10b). Note also that g_0 (≡GW) = 1.0 and RHG = R_L/Gr_L.

A complete listing of the revised computer program for this problem is given in the diskette. The discussion in this section, as well as in Sections 1.4 to 1.6, should be read by printing the computer program and checking its results with the sample calculations of Table 14.1, obtained for uniform wall temperature with Pr = 0.72, R_L/Gr_L = 1, h_1 = 0.16, K = 1 for a total of six ξ-stations.

Table 14.1. Computed wall shear and heat transfer parameters for uniform wall temperature.

$\xi \times 10^3$	f''_w	g'_w
0	0.3320	-0.2956
0.0025	0.4881	-0.3327
0.005	0.4802	-0.3305
0.0075	0.5366	-0.3419
0.01	0.5490	-0.3444
0.015	0.6088	-0.3556

1.4 Computer Program for Wall Jet and Film Heating

A wall jet with fluid temperature higher than that of the freestream is arranged to blow up a vertical, constant temperature wall with a velocity higher than that of the parallel freestream and to describe this problem, we use the computer program of Section 1.3. Since no variational equations are solved, the SOLV5 subroutine contains no extra statements and is the same as that in Section 1.2. The subroutine COEF contains only the definitions pertinent to the standard problem and there is no need to include the statements used for the variational problem.

The only major change occurs in the subroutine IVPL where the initial velocity and temperature profiles are defined according to Eqs. (14.19) and (14.22), that is,

$$u = \begin{cases} 6u_c \left(\frac{\eta}{\eta_c}\right)\left(1 - \frac{\eta}{\eta_c}\right) & 0 \le \eta \le \eta_c \qquad (14.19a) \\ u_e \sin\left[\frac{\pi}{2} \frac{\eta - \eta_c}{\eta_e - \eta_c}\right] & \eta_c \le \eta \le \eta_e \qquad (14.19b) \end{cases}$$

$$g = \frac{1}{2}\,[1 - \tanh\beta(\eta - \eta_c)] \qquad (14.22)$$

The subroutine HER is used to obtain the blending velocity profile between the wall jet and freestream boundary layers according to Eq. (14.20),

$$u(\eta) = u(\eta_1)\psi_1(\eta) + u(\eta_2)\psi_2(\eta) + u'(\eta_1)\bar{\psi}_1(\eta) + u'(\eta_2)\bar{\psi}_2(\eta) \qquad (14.20)$$

where the prime denotes differentiation with respect to η, and

$$\psi_1(\eta) = \left(1 - 2\,\frac{\eta - \eta_1}{\eta_1 - \eta_2}\right)\left(\frac{\eta - \eta_2}{\eta_1 - \eta_2}\right)^2, \quad \bar{\psi}_1(\eta) = (\eta - \eta_1)\left(\frac{\eta - \eta_2}{\eta_1 - \eta_2}\right)^2$$
$$\psi_2(\eta) = \left(1 + 2\,\frac{\eta - \eta_2}{\eta_1 - \eta_2}\right)\left(\frac{\eta_1 - \eta}{\eta_1 - \eta_2}\right)^2, \quad \bar{\psi}_2(\eta) = (\eta - \eta_2)\left(\frac{\eta_1 - \eta}{\eta_1 - \eta_2}\right)^2 \qquad (14.21)$$

To obtain f" and g', we differentiate the above expressions.

A complete listing of the revised computer program for this problem is given in the diskette and Table 14.2 shows the wall shear and heat transfer parameters for a laminar wall jet on a vertical flat plate for Pr = 0.72, Ri = 0.10 and u_c/u_e = 0.50. The calculations were started at z = 1 with initial velocity and temperature profiles as described above. Due to the rapid changes in the profiles, especially in the blending region, a variable grid was used across the layer, as described on p. 441 of the book.

Table 14.2. Computed wall shear and heat transfer parameters for a laminar wall jet on a vertical flat plate.

z	θ'_w	f''_w
1	0	1.0
1.005	-0.00005	0.874
1.01	-0.00009	0.813
1.02	-0.00012	0.793
1.05	-0.00036	0.713
.	.	.
.	.	.
.	.	.
1.35	-0.040	0.503
1.40	-0.050	0.473
1.45	-0.061	0.482
1.50	-0.071	0.458

1.5 Computer Program for Turbulent Free Jet

The computer program of Section 1.2 can also be used to calculate the properties of a two-dimensional, nonsimilar, heated turbulent jet. In this case, the relevant equations with the concepts of eddy viscosity and turbulent Prandtl number are given in transformed variables by Eqs. (14.27) and (14.28) and their boundary conditions by Eq. (14.30).

The linearized finite-difference equations for Eqs. (14.27), (14.28) and (14.30) are given by the system represented by Eqs. (13.62), (13.63) and (14.41) with the turbulence model given by (14.32) The changes to the computer program of Section 1.2 occur in MAIN and in the three subroutines, IVPL, COEF and SOLV5. Subroutine IVPL now has the initial velocity and temperature profiles given by Eq. (14.31),

$$\frac{f'}{3\xi_0^{1/3}} = \theta = \frac{1}{2}\{1 - \tanh\beta\,[\frac{3\xi_0^{2/3}}{\sqrt{R_L}}(\eta - \eta_c)]\} \qquad (14.31)$$

Subroutine COEF contains revised coefficients of the linearized finite-difference equations, $(s_k)_j$, $(\beta_\ell)_j$ and $(r_k)_j$ and MAIN has the eddy-viscosity formulas given by Eq. (14.32). Note that the coefficients of the momentum equation $(s_k)_j$ are identical to those given by Eq. (13.64) provided that we take $m_1 = 1$ and $m_2 = -1$. Except for $(\beta_1)_j$ and $(\beta_2)_j$, the coefficients of the energy equation $(\beta_k)_j$ $(k = 3, \ldots, 10)$ are identical to those given by Eqs. (13.65c) - (13.65j) provided that we take $m_1 = 0$ and $d_j = 0$. The coefficients $(\beta_1)_j$ and $(\beta_2)_j$ are given by Eqs. (14.40). The changes in SOLV5 occur due to the centerline boundary condition of the present problem which requires that $\delta v_0 = 0$ rather than $\delta u_0 = 0$ used in the computer program of Section 1.2. For this reason, the second row of the A_0 matrix of Eq. (13.68a) must be changed to

$$0 \quad 0 \quad 1 \quad 0 \quad 0$$

To incorporate the centerline boundary condition $\delta p_0 = 0$ for the energy equation, we set $\alpha_0 = 0$ and take $\alpha_1 = 1.0$. A complete listing of the revised computer program for this problem with appropriate changes is given in the diskette. Table 14.3 shows the variation of dimensionless centerline velocity and temperature for a turbulent heated jet computed with this computer program for Pr = 0.72 and $R_L = 5300$.

Table 14.3. Variation of dimensionless centerline velocity and temperature for a turbulent heated jet.

ξ	f'_c	g_c
1	3.0	0.999997
1.005	3.005	0.999997
1.01	3.010	0.999997
1.02	3.020	0.999996
1.05	3.049	0.999995
.	.	.
.	.	.
.	.	.
1.50	3.436	0.999388
1.55	3.473	0.999104
1.60	3.509	0.99873
.	.	.
.	.	.
.	.	.
10.0	4.5925	0.692689

1.6 Computer Program for Mixing Layer

The computer program of Section 1.2 can also be used to describe the mixing layer between two uniform streams at different temperatures. For simplicity we consider a laminar similar flow, which requires the solution of Eqs. (8.48), (8.49) subject to the boundary conditions given by (8.44) for specified values of Pr and λ.

To use the program of Section 1.2 with as few changes as possible, we use the nonlinear eigenvalue approach described in Section 14.1 and this requires minor changes in the COEF subroutine. For the standard problem, the boundary conditions have the same form as a boundary-layer flow and the finite-difference coefficients of the momentum and energy equations are identical to those of a uniformly heated flat-plate flow. So we need to set $b_j = 1$, $e_j = 1/Pr$, $d = 0$ and $\alpha_n = 0$ and take $\alpha_0 = 1$, $\alpha_1 = 0$ to handle the boundary conditions of the energy equation. The coefficients of the variational equations also require only minor changes so that, except for the $(r_1)_0$ term which is equal to 1, the right-hand sides of the variational equations $(r_k)_j$ are all zero and their coefficients $(s_k)_j$ and $(\beta_\ell)_j$ are the same as those in the standard problem. As a result, the solution of the variational equations, needs only the recomputation of $\underset{\sim}{r}_j$ and, since the A-matrix in Eq. (13.28) is already known from the standard problem, a solution for $\underset{\sim}{\delta}_j$ which is now defined by Eq. (14.15). Unlike the problem of Section 1.3 (or Section 14.1), however, the calculations in SOLV5 must continue from the "outer" edge $\eta = \eta_j$ up to $\eta = 0$ to determine $(f_1)_{\eta=0}$.

A complete listing of the revised computer program for this problem is included in the diskette together with sample calculations for a mixing layer for which Pr = 0.72 and λ = 0.50. The η-interval consists of $-6 \leq \eta \leq 6$ with h_1 = 0.12, K = 1. Table 14.4 presents velocity and temperature profiles at several selected η-values.

Table 14.4. Velocity and temperature profiles for a mixing layer for λ = 0.50.

η	u	g
-6.0	0.5000	0.1091×10^{-4}
-3.0	0.5244	0.8070×10^{-1}
-1.08	0.6418	0.3196
0.0	0.7649	0.5327
1.08	0.8873	0.7449
2.04	0.9583	0.8831
3.00	0.9894	0.9589
5.04	0.9999	0.9985
6.00	1.0000	1.0000